EXPERIMENTS WITH GENE FUSIONS

Thomas J. Silhavy
NCI-Frederick Cancer Research Facility

Michael L. Berman
NCI-Frederick Cancer Research Facility

Lynn W. Enquist
Molecular Genetics, Inc.

Cold Spring Harbor Laboratory
1984

The cover displays a pencil drawing of the Chimera, by Nancy Trun. In Greek mythology Chimera is a fire-breathing she-monster, commonly represented with a lion's and goat's head, a goat's or lion's body, and a serpent's tail. In biology a chimera is a tissue composed of cells from at least two genetically distinct origins. In molecular genetics the term chimera has come to refer to: (1) a gene composed of two or more genetically distinct loci; specifically, a gene composed of the 5' sequences from one gene fused with the 3' sequences from a second gene (often the 3' sequences from the *lacZ* gene of *Escherichia coli*); (2) a gene product having aminoterminal residues of one peptide and carboxyterminal residues from a second peptide.

Experiments with Gene Fusions
All rights reserved
© 1984 by Cold Spring Harbor Laboratory
Printed in the United States of America
Book and cover design by Emily Harste

Library of Congress Cataloging in Publication Data

Silhavy, Thomas J.
  Experiments with gene fusions.

  Bibliography: p.
  Includes Index.
    1. Gene fusion—Experiments. 2. Gene fusion—
Laboratory manuals. 3. Escherichia coli—Genetics—
Experiments. 4. Escherichia coli—Genetics—Laboratory
manuals. 5. Recombinant DNA—Experiments. 6. Re-
combinant DNA—Laboratory manuals. I. Berman,
Michael L. II. Enquist, L. W. (Lynn W.) III. Cold
Spring Harbor Laboratory. IV. Title.
QH462.G46S54 1984   574.87'3282'0724   83-15230
ISBN 0-87969-163-8

Other manuals available from Cold Spring Harbor Laboratory

Molecular Cloning
Advanced Bacterial Genetics
Hybridoma Techniques
Methods in Yeast Genetics
Experiments in Molecular Genetics
Experiments with Normal and Transformed Cells

All Cold Spring Harbor Laboratory publications are available through booksellers or may be ordered directly from Cold Spring Harbor Laboratory, Box 100, Cold Spring Harbor, New York 11724.

SAN 203-6185

To our wives, Daileen, Marta, and Kathy, for their encouragement, understanding, and continued support.

Preface

This manual is designed to demonstrate the use of gene fusions, transposable elements, and methods of recombinant DNA for genetic analysis in *Escherichia coli*. It consists of experiments and procedures used in the Advanced Bacterial Genetics course offered at Cold Spring Harbor Laboratory during the summers from 1981 to 1983. Through this manual, gene fusion technology becomes accessible to students and scientists who are familiar with the principles and basic manipulations of bacterial genetics as described in J.H. Miller's book, *Experiments in Molecular Genetics* (1972). Such a person can grasp the concepts described herein and perform all of the experiments in three weeks, provided he or she is willing to work in the lab 12–14 hours a day. This manual is not meant to be a compendium of recombinant DNA methods. Scientists, particularly those working with eukaryotic cells, who wish to learn this technology in more depth should consult *Molecular Cloning: A Laboratory Manual* by T. Maniatis, E.F. Fritsch, and J. Sambrook (1982).

Many people have contributed substantially to the production of this manual. Our teaching assistants—Scott Emr, Dolores Jackson, Ronald Taylor, Spencer Benson, Erhard Bremer, Stephen Garrett, and Susan Bear—gave many hours of their time, provided valuable advice, and helped us in ways too numerous to mention. The success of the course, and consequently of the manual, is a direct result of their enthusiasm and hard work. Special thanks must be given to Stephen Garrett, who provided insight into the genetics of *envZ* and constructed many of the strains used in this manual, and to Susan Bear, who by working with us for three years was able to give the course a special continuity and offer helpful suggestions for its improvement.

We thank the many colleagues who have generously provided us with unpublished experimental protocols. We have tried to reference the original sources for specific methods, and we apologize for any errors or omissions.

Over the past three years, Sylvia Lucas has given many hours to the production of this manual. She did all the typing and layout for the earlier versions, reminded us of our deadlines, and helped in the preparation of the manuscript for this edition. Without her, this book would still be in loose-leaf form. During the past year, as we were preparing the manual for publication, Lori Jenkins provided much-needed, additional secretarial support. Doug Owen deserves special thanks for his patience and editorial skills. He and Nancy Ford, Director of Cold Spring Harbor's Publications Department, shared their expertise to add coherence to the book and simplify the process of publication. We are also grateful to Emily Harste for the typographic design of the book and cover.

We are particularly indebted to Jim Hicks and Jim Watson for their continued support of bacterial genetics and for their aid and encouragement to us during our tenure as instructors of the Advanced Bacterial Genetics course.

The success of any course is perhaps best judged by its students. Each of the former students listed below has helped to ensure that the techniques described in the manual actually work. We hope they have profited as much from this experience as we have.

One of these former students contributed her artistic talents, as well; we thank Nancy Trun for her rendering of Chimera, which appears on the cover of this manual.

Finally, we would like to acknowledge the magnanimous support given us by our respective institutions and employers. T.J.S. and M.L.B. were supported in part by the National Cancer Institute, DHHS, under contract NO1-CO-23909 with Litton Bionetics, Inc.

Thomas J. Silhavy

Michael L. Berman

Lynn W. Enquist

Maria-Eugenia Armengod
Peggy Arps
David Baird
Robert Belas
Mary Anne Berberich
David Boxer
Bernard Brownstein
Robert Clarke
James Coulton
Linda DeVeaux
Karen Downs
Daniel Eichinger
Stephen Fahnestock
Thomas Fekete
Robert Franco
Sabine Freundlieb
Alphonse Garcia
Stephen Goff
Susan Hoiseth
Monica Hollstein
Sherman Hom
Bjarne Hove-Jensen
Robin Jones
Heidi Kaplan

Susan Kellman
Roger Levesque
Timothy Lohman
Stanley Maloy
William Marcotte
Warren Masker
Joseph Meier
Penelope Nazos
Dennis Ohman
Ann Progulske
Linda Reha-Krantz
James Ruether
James Ryan
Roger Sanders
Herbert Schweizer
Jyoti Sen
Claire Shurvinton
Hee-Sup Shin
Nancy Trun
Eino Väisänen
Fred Warren
Susan Whorisky
Paul Wolfe
Emanuel Yakobson

Contents

Preface, v
Bacterial Strains, xi
Bacteriophages, xiv

Introduction 1

EXPERIMENTS

Experiment 1	Isolating *lacZ* Fusions by Genetic Transposition	7
Experiment 2	Constructing LacZ⁺ Protein Fusions on Plasmids by In Vitro Mutagenesis or Nonhomologous Recombination	18
Experiment 3	Cloning *lacZ* Fusions on a High-copy-number Plasmid	28
Experiment 4	Analyzing Chromosome Structure by Using Gene Fusions and DNA Hybridization	33
Experiment 5	Constructing λ Transducing Phages by Using Recombinant DNA Techniques	39
Experiment 6	Identifying λ Transducing Phages by DNA Hybridization	43
Experiment 7	Identifying λ Transducing Phages by Genetic Complementation	47
Experiment 8	Isolating Tn*10* Insertions in or near a Gene	53
Experiment 9	Targeted Mutagenesis of the Chromosome	59
Experiment 10	Isolating Chromosomal Deletion Mutations	63
Experiment 11	Isolating Deletion Mutations on a λ Transducing Phage	71
Experiment 12	Targeted Mutagenesis of a λ Transducing Phage	75
Experiment 13	Constructing a Genetic Map	79
Experiment 14	Determining Gene Orientation by Using Gene Fusions to Isolate Specialized Transducing Phages In Vivo	83

PROCEDURES

Techniques

Procedure 1	Preparation of 2-ml High-titer λ Liquid Lysates	89
Procedure 2	Preparation of Phage Plate Stocks	91
Procedure 3	Preparation of 1-liter λ Lysates	93
Procedure 4	Rapid Method for Purifying Phage from Plate Stocks or Small Liquid Lysates	95
Procedure 5	Red-plaque Test for λ *int* and *xis* Functions	97
Procedure 6	Selection of λ Lysogens	99
Procedure 7	Induction of λ Lysogens by Ultraviolet Light	102
Procedure 8	Scoring LacZ$^+$ Phage Plaques with Xgal	104
Procedure 9	Detection of Phage Genes in Prophage Deletions	105

P1 Techniques

Procedure 10	Preparation of P1*vir* Lysates	107
Procedure 11	Preparation of a P1Tn*9clr*100 Lysate	109
Procedure 12	Genetic Transduction Using P1*vir*	111

Gene Fusion Techniques

Procedure 13	Preparation of MudI(*lac*, Ap) Lysates	113
Procedure 14	Transduction with MudI(*lac*, Ap)	114
Procedure 15	Conversion of a MudI(*lac*, Ap) Lysogen to a λ Lysogen	115
Procedure 16	Conversion of a λcI$^+$ Lysogen to a λcIts857 Lysogen	117

Mutagenesis Protocols

Procedure 17	Transfer of Tn*10* from λNK561 to the *Escherichia coli* Chromosome and Preparation of a Random Tn*10* Pool	119
Procedure 18	Isolation of Chromosomal Deletion Mutants Following λ Induction	121
Procedure 19	Selection of Deletion Mutants of λ by Using EDTA Plates	123
Procedure 20	Isolation of Deletion Mutations in Phage Containing a *Dam* Allele	127
Procedure 21	Nitrosoguanidine Mutagenesis	129
Procedure 22	Mutagenesis of λ by Ultraviolet Light	131
Procedure 23	Hydroxylamine Mutagenesis of Phage	133
Procedure 24	*mutD* Mutagenesis of λ	135

DNA Preparations

Procedure 25	DNA Extraction from Bacterial Cells	137
Procedure 26	Large-scale Isolation of λ DNA	140
Procedure 27	Rapid Isolation of λ DNA	142
Procedure 28	Large-scale Isolation of Plasmid DNA	144
Procedure 29	Methods for Rapid Plasmid DNA Isolation	147

Cloning Protocols

Procedure 30	Hybrid Formation	152
Procedure 31	Plasmid Hybrid Formation	154
Procedure 32	Transduction with a λ Library	156
Procedure 33	Selecting Hybrid Phage According to Density in Cesium Chloride	160
Procedure 34	Preparation of Sau 3A Partial Digests of Bacterial DNA	162
Procedure 35	Using Helper Phages to Construct Lysogens of λ D69 Hybrids	166
Procedure 36	Selecting λ Hybrid Phages Constructed with λDamsrIλ3	167

DNA Transfer Protocols

Procedure 37	Transformation with Cells Treated with Calcium Chloride	169
Procedure 38	Transfection of Cells Treated with Calcium Chloride	171
Procedure 39	In Vitro Packaging of λ DNA	173

DNA Biochemistry

Procedure 40	Phenol/Chloroform Extraction of DNA Samples	177
Procedure 41	Ethanol Precipitation of DNA	180
Procedure 42	Drop Dialysis of DNA Preparations	182
Procedure 43	Restriction Endonuclease Digestion and Gel Electrophoresis of DNA	183
Procedure 44	Southern Blot Transfer of DNA	186
Procedure 45	Dehydration of Agarose Gels for DNA-DNA Hybridization	189
Procedure 46	DNA-DNA Hybridization	191
Procedure 47	Plaque Hybridization	196
Procedure 48	Nick Translation of DNA	199
Procedure 49	DNA Purification with BND-Cellulose	201
Procedure 50	Preparation of Mini-gel Filtration Columns	203
Procedure 51	Nuclease BAL-31 Digestion	205

Protein Biochemistry

Procedure 52	Preparation of SDS Protein Extracts	208
Procedure 53	Electrophoresis of Proteins	209
Procedure 54	Maxicells: Protein Expression from Plasmids	213

APPENDIXES

Appendix A	Media and Standard Solutions	217
Appendix B	Phenotypes and Genotypes of *Escherichia coli*	223
Appendix C	Transposable Genetic Elements	226
Appendix D	Notes on Growth and Storage of *Escherichia coli*	231

Contents

Appendix E	Notes on Growth and Storage of λ	**233**
Appendix F	Phenotypes and Genotypes of λ	**236**
Appendix G	Titering Phage Lysates	**239**
Appendix H	Spontaneous Induction and Release of Phage from λ Lysogens	**241**
Appendix I	Cloning Vectors	**244**
Appendix J	Moving Mutations from One Replicon to Another by Recombination	**253**
Appendix K	Genetic Verification of Gene Fusions	**259**
Appendix L	Using λ*plac*Mu1	**261**
Appendix M	The Lactose Operon	**266**
Appendix N	The Omp Regulon	**283**
Appendix O	The Maltose Regulon	**286**
Appendix P	The Arabinose Regulon	**288**

References, 291
Index, 299

Bacterial Strains

Strain[a]	Genotype[b]
BHB2688	F$^-$ recA λ^r (λ Eam4 b2 red3 imm434 cIts Sam7)
BHB2690	F$^-$ recA λ^r (λ Dam15 b2 red3 imm434 cIts Sam7)
KLF41	F'141/leuB6 hisG1 recA1 argG6 metB1 lacY1 gal-6 xyl-7 mtl-2 malA1 rpsL104 tonA tsx supE44
LE30	F$^-$ mutD5 rpsL azi galU95
LE292	HfrH argEam rpoB galT :: (λΔ[int-FII])
LE392	F$^-$ supF supE hsdR galK trpR metB lacY tonA
LE392.23	LE392 Δ(argF-lac)U169
MAL103	F$^-$ Δ(gpt-proAB-argF-lac)XIII rpsL [MudI(lac, Ap)] (Mucts62)
MB100	MC4100 ara$^+$ leu ABCD::Tn10
MB101	MBM7014 Φ(araBA'-lacZ$^+$)101[λp1(209)]
MBM7007	F$^-$ araCam araD Δ(argF-lac)U169 trpam malBam rpsL relA thi
MBM7014	MBM7007 supF
MBM7060	MBM7014 (λp1048)
MBM7060(pMLB952)	
MC1000	F$^-$ araD139 Δ(araABC-leu)7679 galU galK Δ(lac)X74 rpsL thi
MC1000 (pMLB524)	
MC1000 (pMLB1034)	
MC4100	F$^-$ araD139 Δ(argF-lac)U169 rpsL150 relA1 flbB5301 deoC1 ptsF25 rbsR
MC4100 (pRT516)	
MH225	MC4100 Φ(ompC'-lacZ$^+$)10-25, [λp1(209)]
MH2101	MH225 ompR101
MH2472	MH225 ompR472

Bacterial Strains

Strain[a]	Genotype[b]
MH513	MC4100 ara$^+$ Φ(ompF'-lacZ$^+$)16-23, [λp1(209)]
MH5101	MH513 ompR101
MH5473	MH513 envZ473
MH760	MC4100 ompR472
MH1160	MC4100 ompR101
MH1471	MC4100 envZ473
N3098	lig7ts supF
RT3	MC4100 envZ3
RT203	MH225 envZ3
SE3001	MC4100 Δ(malK-lamB)1
SE5000	MC4100 recA56
SG158(pRT516.101)[c]	MC4100 Φ(malP'-lacZ::kan1081.1)1 [λp1(209)cIts857]
SG263	MBM7014 malPQ::Tn10
SG265	F$^-$ Δ(gpt-proAB-argF-lac)XIII ara argE am gyrA rpoB thi supP (P1cry)
SG404	F'141/MC4100 asd (P1cam)
SG480[d]	MC4100 Δ(malPQ-bioH-ompB)61
SG608	MH225 (λpRT2.3)
SG624	MH225 envZ22
SG626	MH225 aroB
SV101	MC4100 malPQ::Tn10
SW101	F$^-$ araD139 Δ(araABC-leu)7679 zab::Tn10 Δ(argF-lac)U169 rpsL150 relA1 flbB5301 deoC1 ptsF25 rbsR
TK821	MC4100 ompR331::Tn10
TK827	MH513 ompR331::Tn10
594	rpsL

[a]All strains are *Escherichia coli* K-12.

[b]Fusions constructed as described in Experiment 1 (p. 7) or Appendix L (p. 261) of the manual contain a λ prophage that lies adjacent to the fusion in the chromosome. This is designated here as [λp1(209)]. Although the fusion phage is a derivative of λp1(209), it is not this phage per se. The event that generated the particular fusion, insertion or deletion, altered the bacterial DNA carried by λp1(209). As a consequence, the phage is "locked in" the chromosome (see Appendix H, p. 241).

[c]See Experiment 10, Fig. 14 (p. 67).

[d]See Experiment 8, Fig. 12 (p. 56).

Other bacterial strains

Enterobacter cloacae

Klebsiella pneumoniae

Proteus mirabilis

Salmonella typhimurium LT2

Serratia marcesens

Shigella sp.

Bacteriophages

Bacteriophage	Genotype[a]
B10	λ imm21 cI
B17	λ int6 red3 imm21 cI
B500	λ h80 imm21 c
G6	λ imm434 cI
G216	λ b2 imm434 cIts
G244	λ b538 imm434 cI Sam7
Y1	λ cIts857 cI ind
Y2	λ b2 cIts857
Y47	λ cIts857 Sam7
Y2223	λ Wam403 cIts857
W14[b]	λ $v_2 v_1 v_3$
W30	λ b2 cI
W248	λ h80 Δ(att-int)9 cI
λNK561	λ b22I cI::Tn10 Oam29 Pam80
λp1(209)[c]	
λplacMu1[d]	
λpMu507	λ cIts857 Sam7 MuA$^+$B'
λp10-25	λ Φ(ompC'-lacZ$^+$)10-25
λp16-13	λ Φ(ompF'-lacZ$^+$)16-13
λTK10	λ Φ(ompR'-'lacZ)hyb1
λpSG1	λ p1(209)lacY::Tn9
λD69	λ bamλ1° Δ(srIλ1-srIλ2) imm21 nin5 shn6°
λpRT2	λD69 bamλ3::(envZ$^+$)ompR$^+$
λpSG10	λD69 bamλ3::ompR$^+$

Bacteriophage	Genotype[a]
λpSG11	λpSG10 h80
λpSG517[e]	λpRT2 Δ(ompR)517
λpRT2-80	λpRT2 h80
λpRT2.3	λpRT2 envZ3
λpRT2.101	λpRT2 ompR101
λDamsrIλ3	λ Dam15 b538 cIts857 srIλ4° nin5 srIλ5°
λpRT1imm434	λ Dam srIλ3::(ompR⁺) imm434
λNF1955[f]	
λp1048[g]	λNF1955 Φ(tryT'-lacY⁺)1048
λp1081.1[h]	
λapmalB13[i]	λ malG⁺F⁺E⁺K⁺lamB' h80 cIts857 Sam7
hy2[j]	λ hPA-2 immλ vir
K20[k]	
MudI(lac, Ap)[l]	
P1vir	P1 vir
P1cam	P1::Tn9clr100
φ80vir	φ80 vir
φ80pSuIII	φ80 psupF⁺

[a] Unless otherwise stated, all fusion transducing phages are derivatives of λp1(209) (see Experiment 3, Fig. 7, p. 30).
[b] λvir.
[c] See Experiment 3, Fig. 7 (p. 30).
[d] See Appendix L, Fig. 28 (p. 262).
[e] See Experiment 13, Fig. 15 (p. 80).
[f] See Appendix I, Fig. 24 (p. 252).
[g] See Experiment 2 and Berman and Jackson (1984).
[h] See Experiment 10, Fig. 13 (p. 65).
[i] See Marchal et al. (1978).
[j] The receptor for hy2 is OmpC.
[k] The receptor for K20 is OmpF.
[l] See Experiment 1, Fig. 1 (p. 8).

Plate 1
Selection of Lac$^+$ mutants on lactose MacConkey agar. This plate shows the result of streaking strain MBM7060 carrying pMLB952 on lactose MacConkey agar as described in Experiment 2, part I. The plate was incubated for 5 days at 37°C.

Plate 2
Mapping *ompR* mutants by using gene fusions and lactose MacConkey agar. A section of a lactose MacConkey plate is shown. A lawn of a strain carrying an *ompR101* mutation that abolishes expression of an *ompC-lac* fusion was spotted with ~100 pfu of four different phages. Clockwise from upper left: (1) λ*p*RT2.101, negative control; (2) λ*p*SG517, a phage that cannot complement *ompR101* but that can recombine with the chromosome to generate *ompR*$^+$; (3) λ*p*RT1*imm*434, which for reasons not fully understood exhibits weak complementation; (4) λ*p*RT2, complementation. See Experiment 13 for details.

Plate 3
Red-plaque test for λ site-specific excision (see Procedure 5). A *mutD*-mutagenized stock of λ *b*515 *b*519 *imm*21 (Enquist and Weisberg 1977) was plated on a lawn of LE292, using TB top agar and galactose TTC plates. The majority of the phage form plaques with red centers (excision proficient). White plaque mutants (excision defective) are seen at a frequency of about 1–2%. The *mutD*-mutagenesis protocol is described in Procedure 24.

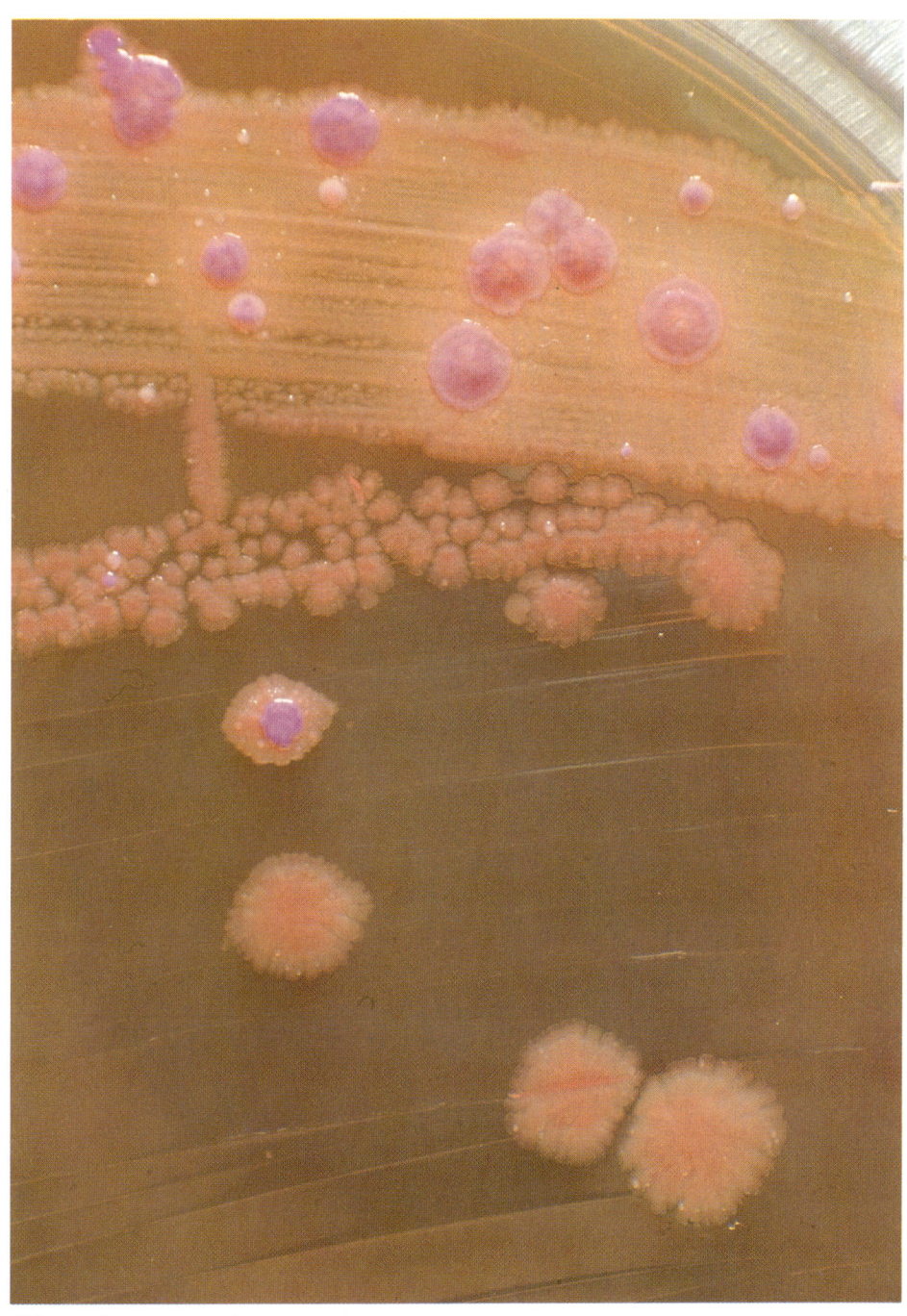

Plate 1

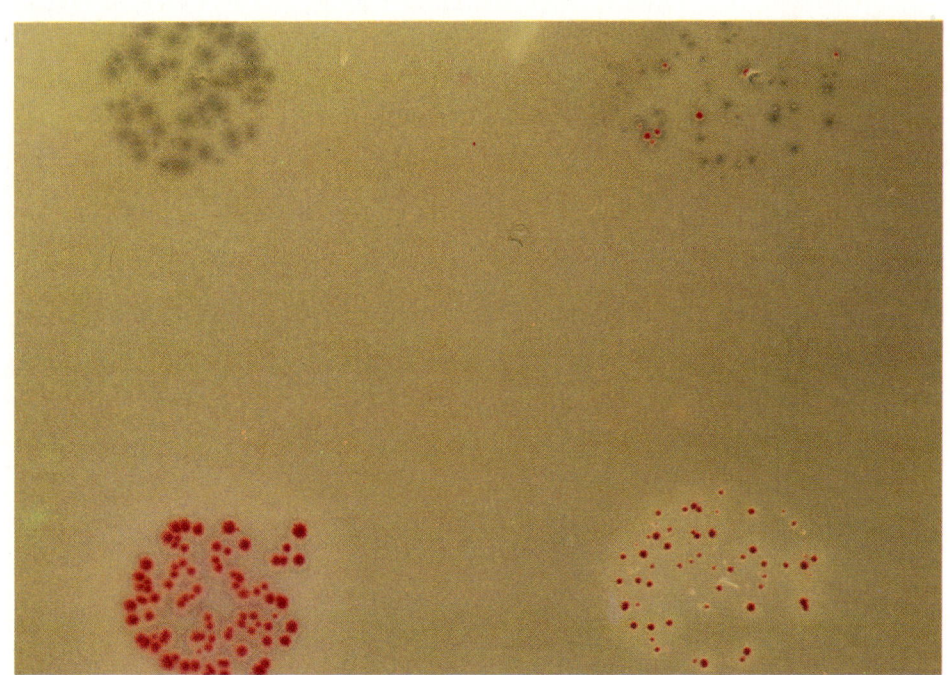

Plate 2

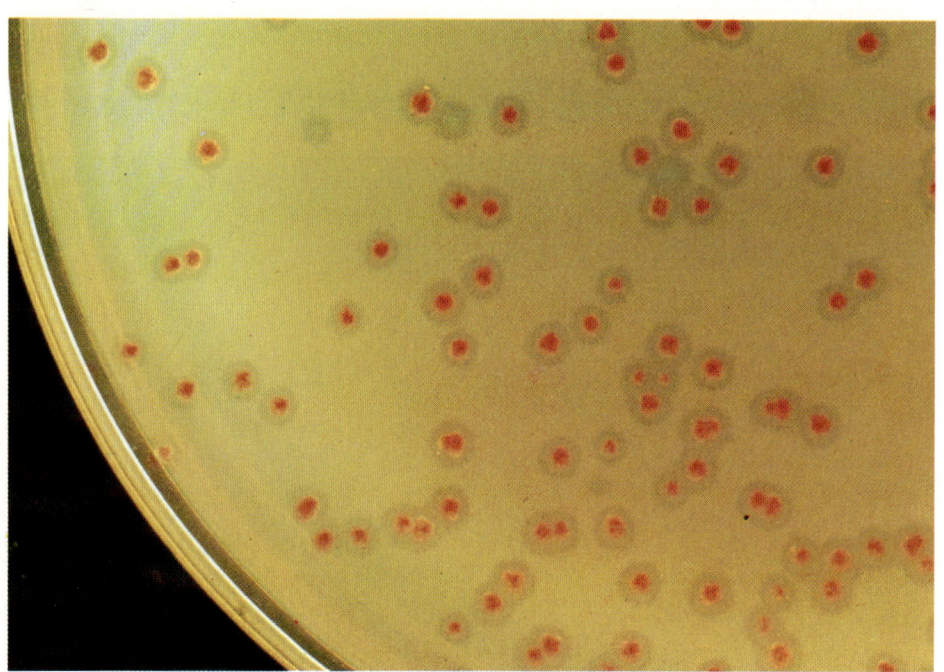

Plate 3

INTRODUCTION

Genetics is one of the most powerful experimental tools available to the biologist. The isolation and characterization of mutants is a rich and productive approach for solving biological problems and formulating scientific concepts. Indeed, the fundamental principles of gene structure, function, and regulation were determined by bacterial geneticists using toothpicks and logic, without the aid of sophisticated instruments and with a minimum of biochemical analysis.

Genetic analysis requires the existence of a selectable or, at least, a scorable phenotype. Without this it is impractical, if not impossible, to distinguish the desired mutant from the overwhelmingly large numbers of wild-type cells present in any given population. Bacterial and bacteriophage genetics succeed simply because environmental conditions can be manipulated to favor growth or to identify, by a color reaction, rare mutant cells or phages as they propagate to form colonies or plaques. This coupled with the extremely rapid growth of bacteria and phages enables one to obtain a large collection of mutants that occur spontaneously at frequencies as low as 1 in 10 billion in a matter of hours. The challenge and the excitement in genetic analysis is in designing methods to isolate these mutants.

One biological system to which genetic analysis has been extensively employed is the lactose (*lac*) operon of *Escherichia coli* (Appendix M). This operon contains two genes whose products are required for growth on lactose. One, *lacZ*, specifies an enzyme, β-galactosidase, that hydrolyzes lactose into its metabolizable components, glucose and galactose. The other, *lacY*, specifies a protein that transports lactose into the cytoplasm. Both gene-product activities are easy to assay. The inherent simplicity of this system permits the formulation of media to detect high, low, intermediate, or no expression of the operon. In addition, a host of synthetic substrates have been made to permit the selection of nearly every conceivable mutant phenotype. In fact, the operon concept was developed as a result of the pioneering genetic studies of lactose utilization in *E. coli* by Jacob, Monod, and co-workers in the 1950s (Jacob and Monod 1961).

Many investigators have long coveted the genetic techniques available with *lac*. The particular systems they studied could not be readily adapted to genetic analysis either because the gene(s) under study were essential or because there were few mutant phenotypes associated with their system and to screen or select for these was difficult. Until recently, these

2   Introduction

problems were often insurmountable. Solutions have now been provided with the discovery of translocatable genetic elements and the advent of recombinant DNA technology. These methods permit one to fuse either the *lac* operon or the *lacZ* gene itself to any gene in *E. coli*, regardless of that gene's function (Experiments 1, 2, Appendix L). Consequently, one can confer the phenotypes of *lac* to any gene, operon, or regulon and then exploit these phenotypes for detailed genetic analysis.

In the past several years, *lac* gene fusions have been used to study a wide variety of biological phenomena. It has been amply demonstrated that these fusions can be employed to study both transcriptional and translational regulation (Bassford et al. 1978; Weinstock et al. 1983; Baker and Wolf 1984). Along similar lines, *lac* fusions have been used to identify genes whose expression is increased by the presence of certain substrates, physical agents such as UV radiation, or various types of physiological stress (Kenyon and Walker 1980; Kolodrubetz and Schlief 1981; Wanner et al. 1981; Strauch et al. 1984). Fusions provide a mechanism to clone regulatory genes (Experiment 7) and to isolate (Experiments 8, 9, 10, 11) and characterize mutations (Experiment 13) that alter regulatory components or *cis*-acting regulatory sites. Since most fusions isolated in vivo contain a λ prophage integrated into the chromosome adjacent to the fusion, specialized transducing phages that carry the fusion and nearby genes can be isolated simply (Experiment 1). Such phages can be used for fine-structure genetic mapping (Experiment 12) and for determining the direction of transcription of the gene to which *lac* is fused (Experiment 14). Techniques for subcloning fusions from the transducing phages onto multicopy plasmid vectors have been described (Experiment 3), and these provide sufficient amounts of fusion DNA and the associated regulatory sites for physical analyses (Experiment 4) and DNA sequencing. Moreover, segments of gene fusions can be used to probe a recombinant DNA library (Experiment 5) to identify particular clones (Experiment 6). In this way, genes can be purified even if their functions are not known.

Certain fusions result in the formation of a hybrid gene that specifies a hybrid protein, composed of aminoterminal sequences from the target gene and a large, functional carboxyterminal portion of β-galactosidase (Experiment 2). Such fusions provide a means to label covalently protein sequences of interest with an enzyme activity that is simple to assay. Using this enzyme activity, one can purify the hybrid protein without difficulty. Antisera raised against the hybrid molecule will cross-react with sequences derived from the gene to which *lac* is fused, thus providing a means to identify and purify the target gene product (Shuman et al. 1980). Moreover, fusions that result in the production of a hybrid protein often retain properties characteristic of the target gene product and, as such, may confer novel phenotypes. These phenotypes have been exploited to study biological problems such as protein-protein interactions (Müller-Hill et al. 1976; Shuman and Silhavy 1981) and mechanisms by which envelope proteins are exported from the cytoplasm (Silhavy et al. 1983).

Despite the proven usefulness provided by fusion technology, many investigators have refrained from applying these methods to their particular experimental systems. Often this is because they are not familiar with the genetic manipulations required to construct and verify gene fusions. Other times, they feel that the gene(s) under their study may not be amenable to these techniques since they may code for essential products and/or the associated mutant phenotypes are limited. This manual is designed to make gene fusion technology accessible to scientists familiar with the principles and basic manipulations of bacterial genetics. Hopefully, it demonstrates the power and utility of using the combined approaches of gene fusion, recombinant DNA methods, and more-classical genetic techniques for the analysis of an experimental system. For the sake of clarity, the experiments described focus on a particular regulon; however, the techniques presented are sufficiently general to be applicable to any gene in *E. coli* for which there exists a mutation conferring a recognizable phenotype.

This manual is divided into three major sections, not counting the introductory material, which includes the genotypes of all bacterial strains and phages used. The first section describes 14 experiments that we have used in the Advanced Bacterial Genetics course at Cold Spring Harbor Laboratory. In writing these experiments, we have tried first to give a general introduction that provides sufficient background for understanding the experimental design. The general purpose and the specific goals of the experiment are stated, followed by the day-to-day schedule of experimental manipulations. Under the heading Specific Comments, we have raised some important points for thought or discussion. These have been collected from students at the Cold Spring Harbor course over the previous three summers. They may include technical problems, subtle points requiring further emphasis, or additional experiments necessary to verify the results obtained.

The second and largest section of the manual is devoted to experimental protocols. These are divided into related groups. They range from basic methods in bacterial and phage genetics through techniques of recombinant DNA to elementary protein biochemistry. As a group, they probably represent most of the techniques currently in use in our collective laboratories. Each of these protocols have been used successfully many times, and we feel confident that they will be of general utility. To facilitate use of this manual, these procedures are listed both in the Contents and in the Index.

The appendixes, the final section of the manual, contain recipes for various media and our recommendations for the care and feeding of *E. coli* and its bacteriophage λ. In addition, they contain concise summaries of the operons, regulons, transposons, and cloning vectors encountered in the manual. Relevant genetic maps and DNA sequences are included, and important genetic techniques and concepts are discussed.

To those who hope to use this manual as a teaching aid, we offer several additional comments. First, it has been our experience that students work best in groups of two. However, to save materials, experiments utilizing more-complicated biochemical techniques were done in groups of four. Second, none of the experiments require sophisticated equipment, provided that instructors prepare certain purified vectors and the in vitro packaging extracts. It does, however, require a lot of media. We consumed approximately 11,000 plates per course, i.e., about 700 plates per student. Preparing all of that media is something none of our assistants will soon forget. Perhaps our most important advice is to stress control experiments. We purposefully omitted many controls because, by requiring the students to do this independently, we could determine quickly who was thinking and who was not. At first we were amazed; the students were doing little more than what was actually written. By the end of each course, however, they were doing controls we had not even thought of. We are confident that none of them will ever assume a strain to be a λ lysogen because it is phenotypically resistant to λcI without also showing that it is sensitive to λ*vir*.

EXPERIMENTS

EXPERIMENT 1
Isolating *lacZ* Fusions by Genetic Transposition

The experiments and techniques described in this manual take advantage of gene fusions to accomplish genetic analysis. This approach extends the range of available phenotypes for any target gene (Bassford et al. 1978; Weinstock et al. 1983). In particular, *lac* gene fusions allow the sophisticated techniques used for *lac* genetics to be applied to any target gene regardless of its function. Convenient features of the *lac* operon include the simplicity of genetic organization, the ease with which the gene products can be assayed, and the variety of media to select or screen for virtually every conceivable Lac phenotype (see Appendix M, p. 266). The purpose of this experiment is to demonstrate modern methods for the construction and manipulation of *lacZ* fusions in vivo.

The *lac* genes can be fused to the target gene in one of two different ways, depending on the exact location of the fusion joint. Lac⁺ fusions that produce a hybrid protein (protein fusions) are the result of the fusion of the *lacZ* structural gene with the target structural gene. Protein fusions are described in Experiment 2. A second type of fusion (operon fusion) places the *lac* genes under the regulatory control of the promoter of the target gene. Operon fusions do not result in the production of a hybrid protein. The genetic structure of operon fusions is such that the transcription-regulating properties of the promoter in question are reflected in the levels of *lac* expression observed. On the other hand, protein fusions are translated from the ribosome-binding site and the initiation codon of the target gene and thus reflect both the transcriptional and translational activity of the target gene. Consequently, by isolating both operon and protein fusions and comparing the regulation of β-galactosidase activity, it is possible to decide whether the regulation of any gene occurs at the level of transcription or translation.

Historically, the target genes for fusions to *lacZY* were limited. That is, the targets had to reside adjacent to *lac* on the chromosome or be close to a prophage attachment site. In the latter case, a *lac*-transducing phage could be translocated to the appropriate *att* site in a strain carrying a deletion of the wild-type *lac* genes and only then could *lac* be fused to neighboring targets. For example, the *lac* genes that are normally located at 8 min on the chromosome map could be moved to *att*ϕ80 at 27 min by using ϕ80d*lac* transducing phages (Beckwith et al. 1967). The *trp* operon, which is located very near the *att*ϕ80 site, then became the target of elegant gene fusion experiments by Beckwith and co-workers. The use of these *trp-lac* fusions helped to define precisely the regulatory signals of both the *trp* and *lac* operons, thereby demonstrating the utility of fusions as an experimental tool (Miller et al. 1972; Mitchell et al. 1975, 1976; Guarente et al. 1977).

A truly general method for constructing fusions requires that translocation of the *lac* genes not be limited to certain chromosomal loci, as is the case with phages like φ80d*lac*. This lack of site specificity is a characteristic of transposable genetic elements, such as the bacteriophage Mu (see Appendix C, p. 226). Casadaban and Cohen (1979) exploited this property of Mu to construct a defective, specialized transducing phage, Mud I(*lac*, Ap), that allows the translocation of the *lac* genes to virtually any site in the bacterial chromosome (Fig. 1).

Mu and MudI(*lac*, Ap) insert more or less randomly in the chromosome. However, with respect to Mu DNA, the insertion event is specific. Like other transposons, these phages can insert in only one of two opposite orientations; the physical continuity of the Mu DNA is never altered by transposition. Insertion of MudI(*lac*, Ap) into the target gene disrupts

Figure 1
Structure of the MudI (*lac*, Ap) transducing phage. The ends of Mu (▨) are designated *s* and *c*. The *c*-end fragment (~25 kb) contains the genes necessary for transposition, some of the Mu structural genes, and the immunity region, including the gene coding for a temperature-sensitive repressor. In MudI (operon fusion phage) the *s* end is ~200 bp (O'Connor and Malamy 1983) and does not contain any transcription terminators. Adjacent to the *s* end is the *trp-lac* W209 protein fusion. The *trp* sequences ('*trpCBA*', ~1.5 kb) replace *lacPO* and the 5' end of *lacZ* and provide the translation initiation site for *lacZ*. In MudII (protein fusion phage) the *s* end is a 117-bp open-reading frame fused to codon 8 of *lacZ*. Transcription and translation initiation signals are provided by the target gene. Both Mu*d* phages carry a gene specifying β-lactamase (*bla*), which confers resistance to ampicillin (Ap^r). Consequently, lysogens can be selected either as Lac$^+$ or Ap^r transductants.

that gene and abolishes its function. Since the *lac* structural genes, without any signals for expression, are incorporated adjacent to one end of MudI(*lac*, Ap), the insertion can generate a Lac⁺ operon fusion if it occurs in the proper orientation. Since the phage also carries a β-lactamase gene (*bla*), it is possible to obtain operon fusions in a single step by selecting mutants in the target gene that are resistant to ampicillin (Ap^r) and Lac⁺.

In the first part of this experiment, we use MudI(*lac*, Ap) to select operon fusions in vivo. The second part of this experiment deals with the conversion of a MudI(*lac*, Ap) lysogen to a λ lysogen. This has certain advantages for genetic analysis as well as for cloning the fusions, as discussed below. Remember, MudI(*lac*, Ap) yields *operon fusions*. Another Mu*d* phage, MudII(*lac*, Ap), which we will not use, has recently been constructed by Casadaban and Chou (1984) and, using techniques described in this experiment, can be employed for the construction of *protein fusions*.

Lac⁺ Insertions in the ara Operon

The specific goal of this experiment will be to select Lac⁺ insertions that are fusions to the *ara* operon (see Appendix P, p. 288). Our selection is based on the fact that *araD* mutants are sensitive to arabinose (Englesberg et al. 1962). Presumably, such strains do not grow in the presence of L-arabinose due to the accumulation of the intermediate L-ribulose-5-phosphate. (The accumulation of a phosphorylated sugar intermediate often inhibits growth. For example, *galE* mutants are sensitive to the presence of galactose in the medium for this reason.) Therefore, insertions in the *ara* genes necessary for the production of L-ribulose-5-phosphate are resistant to arabinose (Ara^r). Mutations that confer resistance to arabinose usually inactivate one of the other genes in the operon, *araBA* or the positive regulatory gene *araC*. Such mutants can be selected by outgrowth on minimal plates containing arabinose and a second carbon source. Lac⁺ gene fusions to *araC* are constitutive, whereas Lac⁺ fusions to *araB* or *araA* are inducible by arabinose (Casadaban 1976a,b). We will try to select and characterize each type of Lac⁺ *ara-lac* fusion.

λ Fusion Transducing Phage from a Lac⁺ Mud Fusion

Although it is simple to generate Mu*d* fusions, this technology does present some problems. First, Mu*d* fusions are unstable because of the ability of Mu*d* to transpose. This can affect the Lac phenotype and make genetic analysis and mutant selections difficult. This can be demonstrated by streaking a Lac⁻ Mu*d* lysogen on lactose MacConkey agar. After a couple of days you will see hundreds of Lac⁺ papillae. Second, the Mu*d* phage cannot be employed to obtain a specialized transducing phage carrying the entire gene fusion. This precludes certain types of genetic analysis (see Experiments 13, 14, pp. 79, 83; Appendix K, p. 259) and also makes cloning of the desired fusion onto multicopy plasmids more difficult (see Experiment 3, p. 28).

A solution to these problems was suggested by Komeda and Iino (1979). These authors took advantage of another plaque-forming transducing phage isolated by Casadaban, λp1(209), to convert a Mud lysogen to a λ lysogen. This procedure has been improved by incorporating a transposon, Tn9 (see Appendix C, p. 226), into the *lacY* gene present in λp1(209). Lysogens of the resulting phage, λpSG1, can be selected as transductants resistant to chloramphenicol (Cm^r). The use of this phage is outlined in Figure 2. First, the Mud fusion strain is lysogenized with λpSG1 (which, like the original Mud, carries no promoter for the *lac* genes). The next step involves isolating a segregant of this lysogen that has lost the Mud phage. This can result in two classes of λ lysogens after segregation of the Mud prophage. One type will be Lac$^+$ and identical with the original Mud fusion (Fig. 2A). The other type will retain the *lacY*::Tn9 and be Cm^r, Lac$^-$ (Fig. 2B). These λ lysogens can be induced to yield plaque-forming phages that have incorporated the *lacZ* genes with the target promoters by illegitimate excision. Such phages can be isolated and used for genetic analysis or to subclone the gene fusions.

Specific Comments

Two common problems have been encountered in the conversion of Mud to λ (Fig. 2). Both of these are related to the presence of multiple prophages. Fusion strains that carry additional Mud insertions located somewhere else in the chromosome cannot be converted simply. Even though the desired recombination with λpSG1 *has* occurred at one prophage, the presence of an additional Mud prophage will result in very low survival at 42°C. Accordingly, it is prudent to demonstrate that a fusion strain contains only a single Mud prophage before proceeding. In the case of *ara-lac* fusions, this is done as follows. As explained in Procedure 10 (p. 107), a P1 lysate is prepared on strain SW101, which contains Tn10 inserted very near an *ara-leu* deletion. If this lysate is used to transduce (see Procedure 12, p. 111) the fusion strain to resistance to tetracycline (Tc^r), most of the transductants should also acquire the linked deletion, and consequently the fusion and Mud will be lost. If *all* transductants are Ap^r, an additional Mud must be present. The frequency of multiple Mud prophages varies from experiment to experiment. The usual way of dealing with this situation is to transduce the Mud-*lac* fusion into the parental background (MC4100) using phage P1 (see Procedure 12). The transduction can cause additional Mud transpositions to occur; i.e., some of the Lac$^+$ transductants will be fusions to other chromosomal loci and not the target gene. Accordingly, it is necessary to characterize the transductants carefully before proceeding.

A second problem arises with the formation of tandem λpSG1 lysogens. Indeed, if each cell is infected with many phages, tandem lysogens will be quite common. Segregation of Mud from these lysogens will not result in the formation of Cm^s, Lac$^+$, Ap^s, temperature-resistant survivors. Such cells will remain Cm^r and will exhibit a characteristically high level of spontaneously released λpSG1 (see Appendix H, p. 241). The loss of λpSG1 from these tandem lysogens by homologous recombination will occur at a frequency of 1–5%. Accordingly, single lysogens can be recognized by stabbing isolated colonies into a lawn of λ-sensitive cells. Colonies that produce no zone of lysis are single lysogens of λpSG1.

Recently a novel, plaque-forming λ phage has been developed for the construction of gene fusions in vivo (Bremer et al. 1984). This phage, λ*plac*Mu1, and its use are described in detail in Appendix L (p. 261). Basically this is a λ phage that carries both ends of Mu and will transpose to generate fusions in much the same way as Mu*d*(*lac*, Ap). Since fusions constructed with this phage are stable and since λ transducing phages that carry the fusion can be isolated directly, it provides obvious advantages over the Mu*d* phages. The *lac* region carried by λ*plac*Mu1 is missing transcription and translation start signals for *lacZ*. Therefore, insertions in frame with a target gene will yield protein fusions. Another phage, λ*plac*Mu50, can be used in the same manner to isolate operon fusions. The arrangement of Mu and *lac* DNA at the sites of insertion in these phages is identical with that of the phages MudII(*lac*, Ap) and MudI(*lac*, Ap), respectively. (See the legend for Fig. 1 for a comparison of MudI[*lac*, Ap] and MudII[*lac*, Ap].)

There are certain instances in which the techniques described in this experiment and in Appendix L cannot be directly employed for the isolation of fusions in vivo. Since fusions are generated by an insertion, the target gene is disrupted. Accordingly, if the target gene specifies an essential function, special steps must be taken. One solution is to design a strategy similar to that described in Experiment 8 (p. 53) for the isolation of Tn*10* insertions in a gene whose product may be essential. Another solution is to isolate the fusion on a plasmid as described in Experiment 2. Both of these methods employ merodiploids to complement the loss of essential function.

When constructing protein fusions, another problem may be encountered. Certain hybrid proteins are lethal when expressed at high levels. For example, this is quite common with fusions of *lacZ* to the genes that specify exported proteins (Silhavy et al. 1983). One solution to this problem is to isolate the fusion under conditions in which target gene expression is low (Silhavy and Beckwith 1983). Another solution involves the use of a derivative of MudII(*lac*, Ap) that carries a *lacZ* amber mutation. With this phage, expression of the hybrid protein occurs only if a nonsense suppressor is provided (Palva and Silhavy 1984).

12 Experiment 1

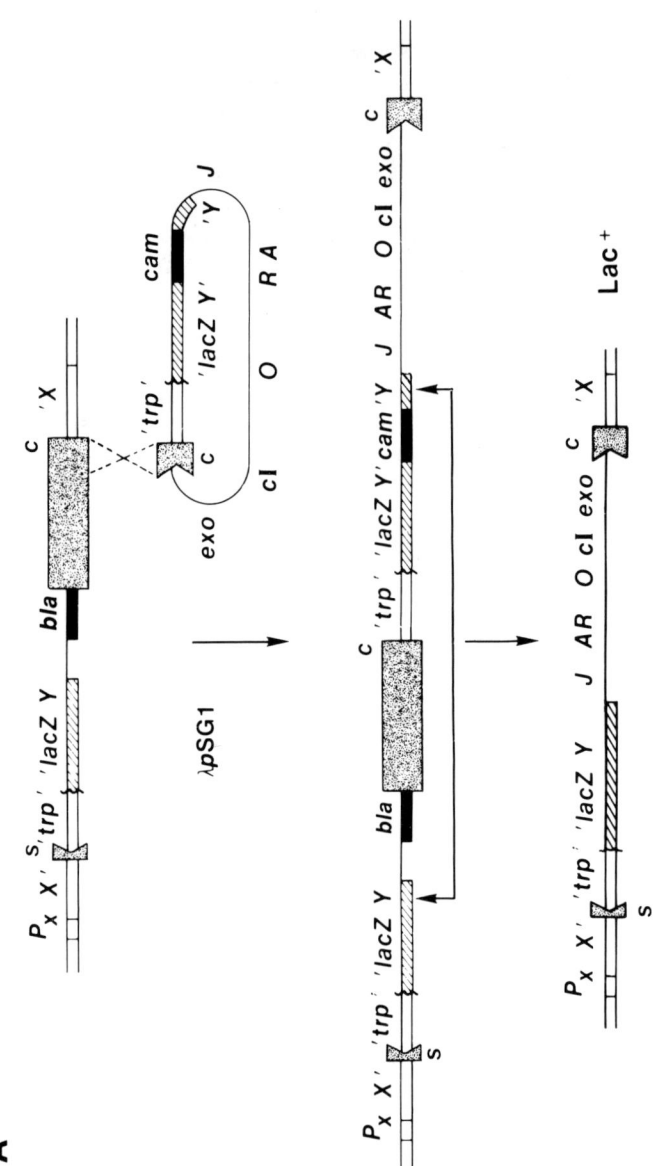

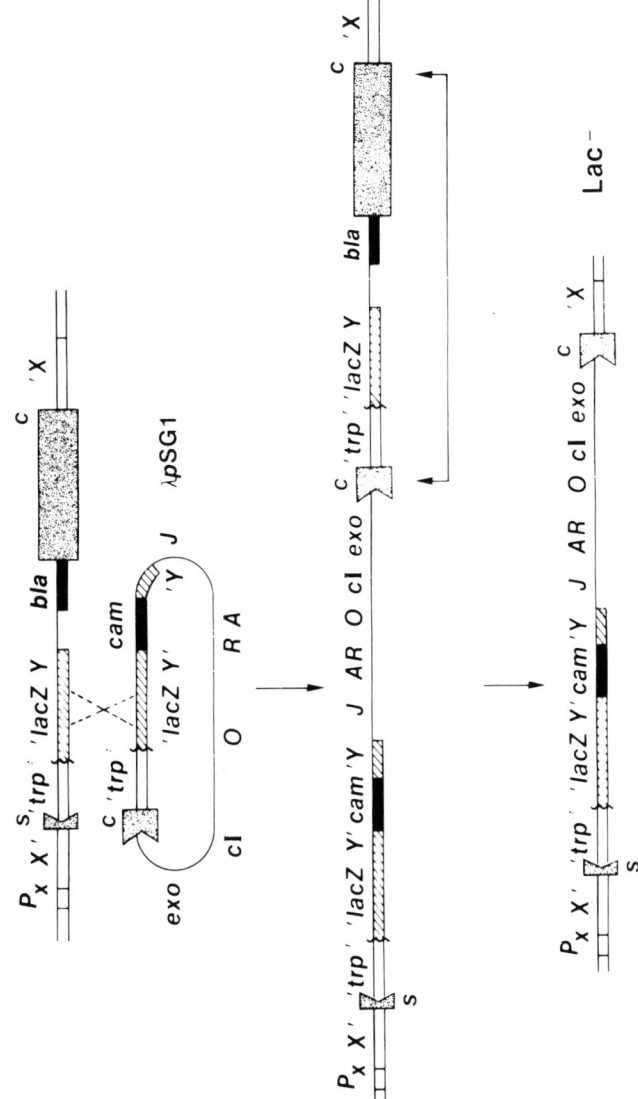

Figure 2

Conversion of a Mud lysogen to a λ lysogen, using λpSG1. A Mud lysogen (Ap^r) is infected with λpSG1, and lysogens are selected as Cm^r. λpSG1 must integrate by *lac* or Mu homology since this phage has no attachment site. (*A*) Integration via Mu homology. This double lysogen (Ap^r, Cm^r) can lose the Mud phage by recombination in *lac* homology. The resulting λ lysogen (Ap^s, Cm^s, Lac$^+$) can be selected according to ability to grow at 42°C since the Mud prophage has a temperature-sensitive repressor. (*B*) Integration via *lacZ* homology. Subsequent loss of the Mud prophage by recombination within Mu homology will result in an Ap^s, Cm^r, Lac$^-$ λ lysogen. To obtain a Lac$^+$ λ lysogen, it is necessary to start with a Lac$^+$ λ/Mud lysogen as illustrated in *A*. It is not possible by homologous recombination to obtain a Lac$^+$ λ lysogen from the Lac$^-$ λ/Mud lysogen illustrated in *B*.

1. Lac⁺ FUSIONS USING MudI(lac, Ap)

DAY 1

Inoculate 5-ml cultures of MC4100 and MAL103. Strain MAL103 must be grown at 30°C.

DAY 2

Prepare Mud lysate (see Procedure 13, p. 79). Centrifuge the MC4100 culture and resuspend the cell pellet in 0.5 volume of 10 mM $MgSO_4$. Do Mud transduction according to Table 1 (see Procedure 14, p. 114).

DAY 3

Replica plate the Ap^r Mud transductants from the L plates onto lactose MacConkey plates containing ampicillin (25 µg/ml). Incubate plates at 30°C. Inoculate a 5-ml culture of SW101.

DAY 4

Record the number of Ap^r transductants and the frequency of Lac⁺ colonies from the lactose MacConkey plates. A good Mud lysate should yield several hundred Ap^r transductants in tubes 3 and 4 (0.1 ml of a 10^{-3} dilution). Note the ratio of Lac⁺, Ap^r to Lac⁻, Ap^r. Is it higher or lower than you expected?

From the Ara^r selection (M63 containing glucose, arabinose, and ampicillin), purify and test the Lac phenotype by streaking at least 24 survivors on two lactose MacConkey plates containing ampicillin (25 µg/ml), one with and one without arabinose (0.1 ml of 20% arabinose stock per plate). Remember to do all this at 30°C. Prepare a P1 lysate on strain SW101 (see Procedure 10, p. 107).

Table 1

Mud Transductions

Test tube	Mud lysate (dilution)	MC4100	Medium
1	0.1 ml (10^{-2})	0.1 ml	L containing 25 µg/ml Ap
2	0.1 ml (10^{-2})	0.1 ml	L containing 25 µg/ml Ap
3	0.1 ml (10^{-3})	0.1 ml	L containing 25 µg/ml Ap
4	0.1 ml (10^{-3})	0.1 ml	L containing 25 µg/ml Ap
5	0.1 ml (undiluted)	—	L containing 25 µg/ml Ap
6	—	0.1 ml	L containing 25 µg/ml Ap
7	0.1 ml (undiluted)	0.1 ml	M63, glucose, arabinose 25 µg/ml Ap
8	0.05 ml (undiluted)	0.1 ml	M63, glucose, arabinose, 25 µg/ml Ap
9	0.1 ml (undiluted) (10^{-2})	0.1 ml	M63, glucose, arabinose, 25 µg/ml Ap
10	0.05 ml (undiluted) (10^{-2})	0.1 ml	M63, glucose, arabinose, 25 µg/ml Ap
11	0.1 ml (undiluted)	—	M63, glucose, arabinose, 25 µg/ml Ap
12	—	0.1 ml	M63, glucose, arabinose, 25 µg/ml Ap

II. ISOLATION OF λ FUSION TRANSDUCING PHAGES

DAY 5

Inoculate 5-ml cultures of the *ara-lac* Mu*d* fusion strains. Grow overnight at 30°C.

DAY 6

Use 2 ml of the overnight cultures of the Mu*d* lysogens as recipient in a P1 transduction using P1*vir* grown on SW101 (see Procedure 12, p. 111). Allow 1–2 hr for expression of the Tc^r phenotype and select Tc^r transductants on L plates containing 25 µg/ml tetracycline.

Isolate λ*p*SG1 lysogens (see Procedure 15, p. 115). After 4–6 hr, streak from phage spots onto lactose MacConkey agar containing chloramphenicol (25 µg/ml). Remember to add 0.1 ml of 20% arabinose to the lactose MacConkey plates if your fusion is inducible by arabinose.

DAY 7

Continue with Day 4 of Procedure 15. Purify Tc^r transductants from Day 6 on lactose MacConkey agar at 30°C. If your fusion is inducible, add arabinose.

DAY 8

Test Cm^r lysogens for *imm*λ (see Procedure 15). Score results after 4–6 hr. Streak λ lysogens at 30°C and 42°C on lactose MacConkey plates. Include Mu*d* parents as well as the original recipient strain, MC4100.

Score the phenotype of the Tc^r transductants from Day 7. Check these transductants for growth at 42°C and for Ap^r.

DAY 9

Using L plates at 37°C, purify temperature-resistant survivors from the Mu*d*/λ lysogens streaked on lactose MacConkey plates.

Score results with Tc^r transductants. If all of these transductants remain Ap^r and temperature sensitive, the fusion strain carries more than one Mu*d* prophage. If this is the case, see Specific Comments.

Experiment 1   17

DAY 10

Test temperature-resistant strains for:

1. retention of λ by cross-streak against λcI (see Procedure 6, p. 99);
2. loss of Mu*d* by streaking on L ampicillin plates (25 μg/ml);
3. loss of Tn*9* by streaking on L chloramphenicol plates (25 μg/ml);
4. Lac phenotype by streaking on lactose MacConkey plates with and without arabinose.

Inoculate a 5-ml culture of MBM7007.

DAY 11

Centrifuge the overnight culture of MBM7007, resuspend in 0.5 volume of 10 mM $MgSO_4$, and store at 4°C.

Score results. Mu*d* fusion strains that have been successfully converted to λ lysogens are Ap^s and temperature resistant. Those that are Lac^+ should be Cm^s. Inoculate 5-ml cultures of these Lac^+ fusions and prepare λ lysates by UV induction (see Procedure 7, p. 102). Titer these lysates, using the MBM7007 cell suspension prepared above on TB medium containing the indicator Xgal and arabinose.

DAY 12

Purify dark blue plaques, using the MBM7007 cell suspension prepared on Day 11, on TB agar containing Xgal and arabinose.

DAY 13

Pick isolated, dark blue plaques from the TB plates containing Xgal and arabinose and prepare lysates (see Procedure 1, p. 89) on MBM7007. Check inducibility by spot titering on TB agar containing Xgal with and without arabinose.

EXPERIMENT 2
Constructing LacZ+ Protein Fusions on Plasmids by In Vitro Mutagenesis or Nonhomologous Recombination

The isolation of gene fusions in vivo (see Experiment 1; Appendix L, p. 261) represents one of the first steps for expanding the genetic analysis of any target gene. However, there are a number of situations in which the in vivo approach is not convenient. These include instances when changes in target gene phenotype are difficult to detect by selection or screening, when the target gene is not an *Escherichia coli* gene, when only a segment of the target gene is available, or when the coding capacity of a cloned DNA segment is not known. In all these cases, techniques of recombinant DNA can be used to facilitate the generation of gene fusions. The general purpose of this experiment is to use these methods to isolate LacZ+ protein fusions, i.e., fusions that create hybrid genes specifying hybrid proteins.

The β-galactosidase monomer has several properties that make the construction and isolation of protein fusions simple and convenient. This was first demonstrated by Müller-Hill and Kania (1974). These authors described an in vivo selection for deletion events that fused *lacI* to *lacZ* and yielded hybrid proteins with β-galactosidase activity. On the chromosome, the *lacI* gene is adjacent to *lacZ*, and both genes are transcribed in the same direction. The order is *lacI*, *lacZ* (see Appendix M, p. 266). One of these hybrids, for example, was shown to be fused between codon 356 of *lacI* and codon 24 of *lacZ* (Brake et al. 1978). The important conclusion was that one can replace the wild-type amino acids at the amino terminus of β-galactosidase with other amino acids (in this case, from the *lac* repressor) and still retain enzyme activity.

Studies of wild-type β-galactosidase, various mutants, and hybrids (Zabin 1982) have shown that the active form of β-galactosidase is a tetramer. Amino acids at the carboxyl terminus play a role in monomer-monomer interaction, while amino acids at the amino terminus play a role in dimer-dimer interaction. Whereas substitutions or deletions of even a few amino acids from the carboxyl terminus will interfere with monomer-monomer interaction, parts of the aminoterminal region can be eliminated or replaced without affecting β-galactosidase activity. Amino acids in the region from residues 26 to 32 seem to be critical for oligomerization. These hydrophobic amino acids must remain intact in order to retain β-galactosidase activity. However, there appears to be no such specificity in the first 25 amino acids of β-galactosidase. Substitutions of these do not affect activity. Functional monomers can be formed by replacing the first 19 to 25 amino acids with as few as 5 (Sarthy et al. 1977) or as many as several hundred new amino acids (Heidecker and Müller-Hill 1977; Emr and Silhavy 1980). The only obvious requirement is that a gene fused to *lacZ* be actively transcribed and provide an in-frame translation start signal.

Protein fusions provide a unique and versatile experimental tool. First, they can be used to study transcriptional or translational control of the target gene (Weinstock et al. 1983). In addition, the purified hybrid protein can be used as an antigen to raise antibodies against the target gene product (Shuman et al. 1980). Moreover, since there appears to be no restriction on the extent of the target gene DNA fused to *lacZ*, the isolation of a series of hybrid LacZ proteins of varying size will provide a mechanism for the construction of a deletion map of any target gene (see Experiment 13, p. 79). Finally, since many hybrid proteins retain all or a portion of the activity of the target gene product (Müller-Hill and Kania 1974; Emr and Silhavy 1980; Berman and Jackson 1984), these fusions can provide a means to identify functional domains within the target gene product.

In addition to our ability to select or screen for LacZ⁺ hybrid genes, it is simple to verify the presence and determine the molecular weight of the hybrid protein. Since the LacZ⁺ sequences present in hybrid proteins are essentially constant, determination of the apparent molecular weight permits a reasonably accurate estimate of the lengths of target gene sequences present in the hybrid gene. The monomer size of 1023 amino acids makes it one of the largest proteins in the cell. On a 7% SDS-polyacrylamide gel, β-galactosidase migrates below the β (m.w. 155,000) and β' (m.w. 165,000) subunits of RNA polymerase in a region of the gel essentially devoid of other proteins. Therefore, hybrid proteins can be simply identified following electrophoresis of whole-cell extracts. Experience has shown that using a Coomassie brilliant blue stain, hybrid proteins can be detected when present at levels as low as 0.5% of cell protein (~750 units of enzyme activity).

This experiment utilizes the plasmid vector pMLB1034 (see Appendix I, p. 244). This vector is a pBR322 derivative in which a part of the *lac* operon replaces the Tcʳ determinant. The presence of this plasmid can be monitored by expression of *bla* (Apʳ). The segment of *lacZ* cloned in pMLB1034 is missing the sites necessary for transcription and translation initiation. In addition, the first 8 amino acid codons of the structural gene are deleted. As described above, all the information to specify an active *lacZ* monomer is contained on this vector. Any DNA insert that correctly aligns translation and transcription start signals with *lacZ* will yield a hybrid gene that confers a LacZ⁺ phenotype[1].

The successful use of this vector requires that one obtain a DNA fragment carrying the 5' end of the target gene. Moreover, the coding

---

[1] Although there is an absolute requirement for an in-frame translation start signal, a bona fide transcription start signal need not always be present. Low-level readthrough transcription from pBR322 promoterlike sequences may be sufficient.

sequences of this gene must be properly aligned with *lacZ*. To make pMLB1034 more useful, multiple cloning sites are incorporated adjacent to the truncated *lacZ* gene. If the DNA sequence of the target gene is known, it is often possible to choose an appropriate restriction fragment by inspection. However, if the DNA sequence is unknown or if no convenient restriction sites are present, construction of the desired hybrid gene by direct insertion can be difficult. We have developed an alternative approach for constructing fusions that does not require such specific information.

The method we use to construct gene fusions is a two-step method. First, a DNA fragment carrying the *entire* target gene is cloned into one of the multiple cloning sites of pMLB1034. To obtain the desired fusion, the transcriptional orientation of the target gene and *lacZ* must be the same. Unless the insertion of this fragment results in the fortuitous formation of a functional hybrid gene, the plasmid will confer a LacZ$^-$, Target Gene$^+$ phenotype. In the second step we identify plasmids that confer a LacZ$^+$ phenotype. A major class of these are plasmids that have suffered a deletion event fusing the target gene to *lacZ*. These will confer a LacZ$^+$, Target Gene$^-$ phenotype. An added advantage of this method, in contrast to the direct insertion approach, is that it allows the isolation of a series of hybrid genes containing varying amounts of target gene sequences (Berman and Jackson 1984). As mentioned above, such a series has numerous useful applications.

The specific goal of this experiment is to isolate a series of gene fusions between *ompR* (see Appendix N, p. 283) and *lacZ*. For this purpose we use a plasmid that we have already constructed, pMLB952 (Fig. 3). This plasmid contains a fragment from pRT516 carrying an intact *ompR* gene and part of *envZ* (Taylor et al. 1983), with the insert oriented such that the direction of transcription is the same as *lacZ*. This plasmid confers a LacZ$^-$ phenotype, indicating that there are no translation start signals in frame with *lacZ*. Any deletion that aligns *lacZ* in frame with either structural gene (*ompR* or *envZ*) can result in the formation of a hybrid gene that specifies a LacZ$^+$ hybrid protein.

This experiment describes two techniques using pMLB952 for obtaining *ompB-lacZ* gene fusions. The first selects for spontaneous in vivo deletions that give rise to a Lac$^+$ phenotype. It requires the use of a particular host strain, MBM7060. This strain is lysogenized with a λ phage that carries a *lacY* operon fusion. In this strain, *lacY* is expressed at high levels constitutively. It is necessary to use this strain for Lac$^+$ selections because plasmid pMLB1034 and its derivatives do not carry an intact *lacY* gene. The second technique utilizes the quasiprocessive exonuclease BAL-31 (Lau and Gray 1979) to generate deletions in vitro. In this case, LacZ$^+$ clones are identified by screening Ap^r transformants with Xgal.

Deletions that create hybrid genes are characterized by performing restriction enzyme digests of plasmid DNA. In addition, we will determine the size of the hybrid protein by SDS-polyacrylamide gel electrophoresis.

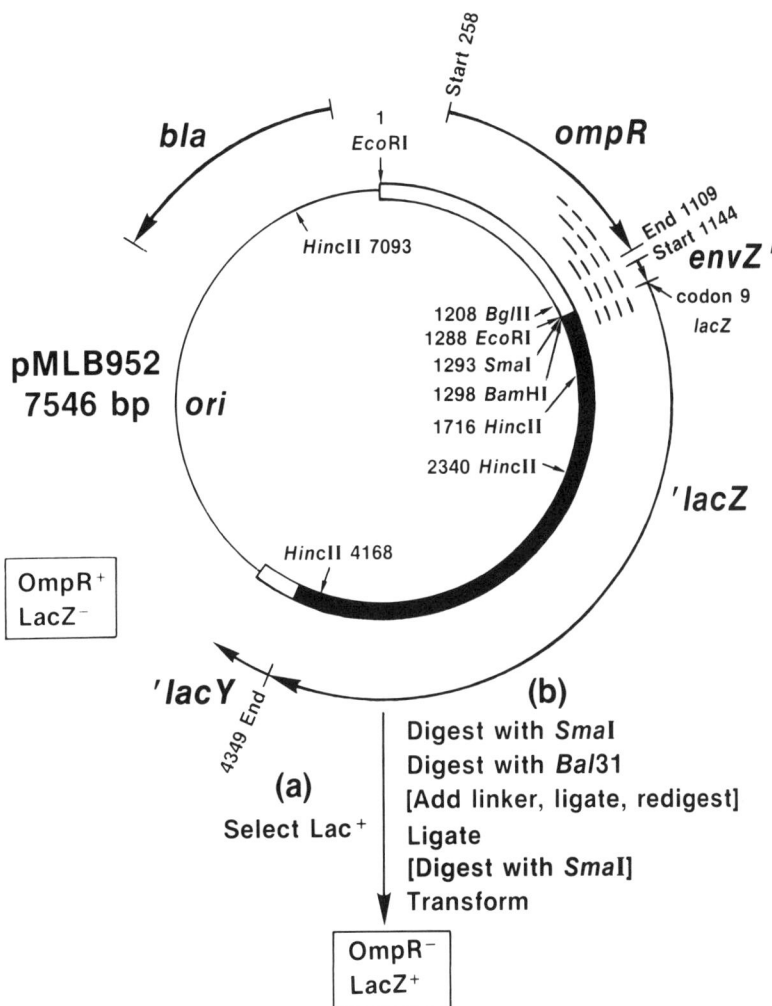

Figure 3
Structure and use of pMLB952 to obtain *ompR-lacZ* protein fusions. This plasmid (Berman and Jackson 1984) is a derivative of pMLB1034 (see Appendix I, p. 244). Relevant structural genes and restriction enzyme recognition sites are indicated. (— — —) Deletions that remove DNA between *ompR* and *lacZ*, resulting in the formation of protein fusions.

Experiment 2

Specific Comments

The plasmid pMLB952 contains the 1288-bp EcoRI-SmaI fragment from pRT516 cloned into the EcoRI site of pMLB1034; the DNA sequence of this fragment has been published (Wurtzel et al. 1982). In pMLB952, the envZ coding sequence is out of frame with lacZ. In the lacZ frame there is a UGA nonsense codon immediately preceding this structural gene at the EcoRI site. The DNA sequence in the vicinity of the EcoRI-SmaI-BamHI linker is as follows (the two-letter amino acid code and the numbering of base pairs is described in Appendix M, Fig. 29, p. 273):

```
     1264       1274       1284       1294       1304       1314
GAAGAGGCAGGTCTGCGTTGGGCGCAACACTATGAATTCCCGGGGATCCCGTCGTTTTACAACGTC
Gu Gu Al Gy Le Ar Tr Al Gn Hi Ty Gu Ph Pr Gy Il Pr Se Ph Ty An Va         EnvZ
                              ** Il Pr Gy Ap Pr Va Va Le Gn Ar Ar         LacZ
```

This explains why plasmid pMLB952 confers a totally LacZ⁻ phenotype. Obviously, deletions that form lacZ fusions with either ompR or envZ must remove this nonsense codon.

This experiment provides a vivid demonstration of the selective power of indicator media, such as MacConkey, that contain high sugar concentrations. As described in Appendix M (p. 266), such media provide a means to screen an enormous number of cells for very rare mutants. Papillae represent the outgrowth of cells able to use the carbon source in the indicator plate. In this experiment, each Lac⁺ papilla within a single colony results from an independent mutational event that occurred during colony formation. Consequently, a large number of independent mutants can be obtained from a single plate.

The frequency of preexisting Lac⁺ mutants in the population can be deduced by plating aliquots of cultures directly on lactose minimal media. Students are encouraged to calculate this frequency and consider it in light of the plasmid copy number and the number of Lac⁺ papillae that appear on lactose MacConkey agar.

The use of BAL-31 to obtain ompR-lacZ fusions suffers from several limitations. First, one of the deletion endpoints must lie within the ompR structural gene. The other endpoint must lie in a very small region of lacZ (between codons 9 and 25). Only a small percentage of BAL-31-treated molecules will have both ends trimmed appropriately. To make matters worse, only 1/9 of these will ligate to restore the correct translational reading frame across the hybrid gene. Finally, the combined inefficiency of DNA ligation and transformation makes it difficult to obtain sufficient numbers of BAL-31-treated plasmids to recover LacZ⁺ clones. Nevertheless, most students are able to demonstrate that almost every BAL-31-treated sample does, in fact, yield deletions of vector DNA. Students should realize that although it is easy to generate deletions using BAL-31, it is often difficult to find specific deletion classes.

Both techniques used for identifying plasmids carrying gene fusions rely on a LacZ⁺ phenotype. This phenotype is dominant. Accordingly, many LacZ⁺ cells will harbor not only the desired plasmid, but also the parental plasmid. This is especially true for the Lac⁺ clones isolated in vivo. Since OmpR⁺ is dominant as well, this precludes scoring the OmpR phenotype immediately. To score this phenotype properly, the mutant plasmid must be introduced into an ompR recipient. Although we do not do this, one method employs the transducing phage λp1048 present at attλ in MBM7060. Since this phage shares lac homology with the Lac⁺ derivatives of pMLB952, a spontaneously released phage (see Appendix H, p. 241) can recombine with the plasmid to form a hybrid replicon (see Appendix J, p. 263). The size of the plasmid and phage are such that this hybrid can be packaged. Consequently, these particles can be recovered by infecting a λ-sensitive, ampicillin-sensitive strain.[2] By employing a *lac* deletion strain as recipient and by using Xgal, we can recognize transductants that acquire a plasmid containing a functional hybrid gene. Students should recognize that plasmid formation in the transductant occurs in the same manner as hybrid replicon formation in the parent strain.

Remember, hybrid proteins can be bifunctional, retaining properties of each of the composite gene products. With *ompR-lacZ* fusions, this may be particularly relevant. If the model for OmpR function described in Appendix N (p. 283) is correct, then we predict that the carboxyl terminus of the protein is not required to activate expression of *ompF*. It would be required only to activate *ompC* expression. Accordingly, it may be possible to isolate *ompR-lacZ* fusions that confer an OmpF$^+$, OmpC$^-$ phenotype in *ompR* strains. If such a fusion can be isolated, it would provide a clear demonstration of how hybrid proteins can be used to study functional protein domains.

---

[2] It is important when doing this transduction to wash the infected cells before plating to remove the large amounts of β-lactamase present in the culture fluid of ampicillin-resistant cells.

24   Experiment 2

I. ISOLATION OF Lac⁺ PROTEIN FUSIONS

METHOD A: *In Vivo Selection of ompR-lacZ Protein Fusions*

DAY 1

Inoculate a 5-ml culture of MBM7060 (pMLB952) in L medium containing ampicillin (125 µg/ml).

DAY 2

Centrifuge the culture and resuspend the cells in 0.5 volume of 10 mM $MgSO_4$. Plate 0.5 ml of the suspension on each of three lactose minimal plates containing ampicillin (125 µg/ml). From the remaining cells, streak the entire surfaces of two lactose MacConkey plates containing ampicillin (125 µg/ml) for single colonies. Incubate all plates at 37°C.

DAYS 3-5

Inspect lactose minimal plates for Lac⁺ colonies and calculate the Lac⁺ mutation frequency. Count the number of Lac⁺ papillae on the lactose MacConkey plates (see Plate 1). Purify 16 Lac⁺ papillae on lactose MacConkey plates containing ampicillin (125 µg/ml). Continue to incubate original lactose MacConkey plates.

DAY 6+

Restreak Lac⁺ colonies on L plates containing ampicillin (125 µg/ml). Prepare plasmid DNA and analyze as described below. Observe that Lac⁺ papillae continue to appear with time.

METHOD B: *In Vitro Construction of ompR-lacZ Protein Fusions*

DAYS 1-2

Obtain pMLB952 DNA that has been purified by banding in $CsCl_2$ (see Procedure 28, p. 144). Digest plasmid DNA with *Bgl*II (see Procedure 43, p. 183). Digest with BAL-31 (see Procedure 51, p. 205).

    Inoculate a 5-ml culture of MC1000 the day before you plan to do plasmid DNA transformation.

DAYS 3–5

Prepare competent MC1000 cells (see Procedure 37, p. 169). Transform these cells with 10 µl of BAL-31–treated DNA that has been ligated. Select transformants on L plates containing ampicillin (125 µg/ml) spread with 0.1 ml of Xgal.

DAY 5+

Purify four white (LacZ$^-$) and any blue (LacZ$^+$) Ap^r transformants by restreaking. Prepare plasmid DNA and analyze as described below.

II. MAPPING THE FUSION JOINT BY RESTRICTION ENZYME ANALYSIS OF PLASMID DNA AND BY ESTIMATING THE SIZE OF THE HYBRID PROTEIN ON SDS-POLYACRYLAMIDE GELS

Inoculate a single colony of each strain from a fresh L plate containing ampicillin (125 µg/ml) into two tubes containing 5 ml of L broth with ampicillin (125 µg/ml). Include the parental strain as a control. Grow the culture until cells reach late-log phase (~10^9 cells/ml). Use one culture for the preparation of cell extracts for SDS-polyacrylamide gel analysis (see Procedure 52, p. 208). Use the remaining culture to prepare plasmid DNA (see Procedure 29A, p. 147). Prepare and run a 7% SDS-polyacrylamide gel (see Procedure 53, p. 209). Prepare analytical *Hinc*II digests of plasmid DNA and analyze on 0.8% agarose gels (see Procedure 43, p. 183).

Examples of results obtained previously with this experiment are shown in Figures 4 and 5.

26   Experiment 2

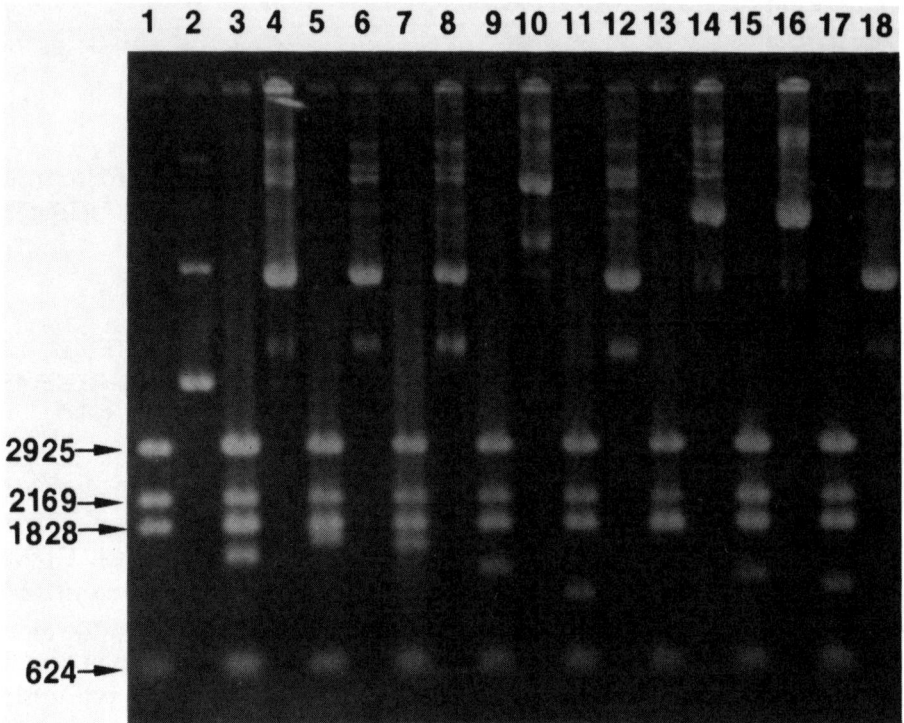

Figure 4
Analysis of plasmids from Lac⁺ mutants selected in vivo. Shown is a photograph of a 0.8% agarose gel stained with ethidium bromide and illuminated by UV light. (Lane *1*) A *Hinc*II digest of purified pMLB952; (lane *2*) the corresponding undigested control; lower band is the supercoiled, monomer-length plasmid, and upper band is relaxed (nicked) pMLB952. (Lanes *3, 5, 7, 9, 11, 13, 15, 17*) *Hinc*II digests of plasmids isolated from Lac⁺ mutants (method A); (lanes *4, 6, 8, 10, 12, 14, 16, 18*) the corresponding undigested controls. The *Hinc*II fragments (in bp) from pMLB952 are 2925, 2169, 1828, and 624. Deletion events, resulting in LacZ⁺ plasmids, will be confined to the 2169-bp fragment. The plasmids isolated from Lac⁺ strains yield the parental size of 2169 bp as well as a unique, smaller DNA fragment. This indicates the presence of at least two plasmids, LacZ⁺ and parental LacZ⁻, in each mutant. Note that the undigested controls show multimer forms of the pMLB952 plasmids in the Lac⁺ strains. In this experiment the Lac⁻ starting strain carried mostly dimer forms of pMLB952 (data not shown). This result was obtained by D. Eichinger and M.E. Armengod.

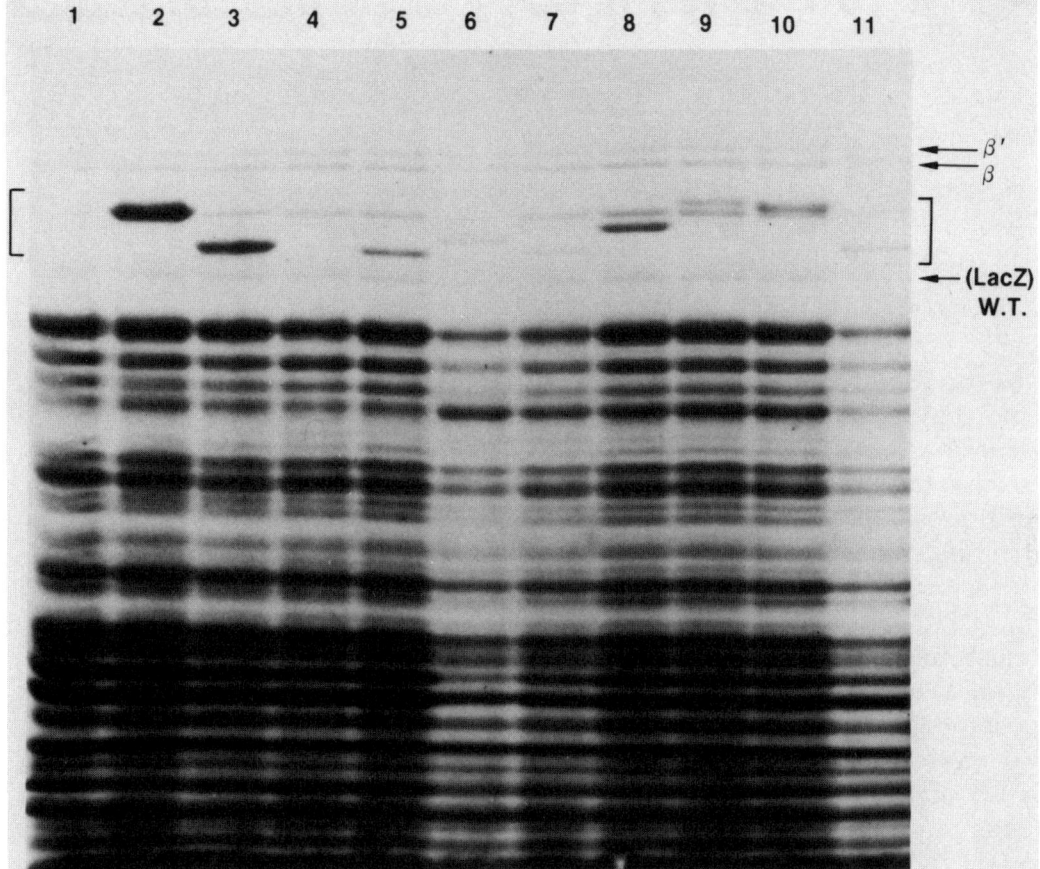

Figure 5
SDS-polyacrylamide gel electrophoresis of whole-cell extracts of Lac⁺ mutants. This is a photograph of a 7% SDS-polyacrylamide gel displaying proteins from various *ompR-lacZ* fusions isolated as described (method A). The cell extracts were prepared according to Procedure 52 (p. 208), and the gel was prepared and developed following Procedure 53 (p. 209). (Lane *1*) Lac⁻ parental strain MBM7060 (pMLB952); (lanes *2–11*) various Lac⁺ mutants. Restriction enzyme analysis of the plasmid DNA isolated from the strains shown in lanes *8–10* is shown in Fig. 4, lanes *3*, *5*, and *7*, respectively. Arrows indicate the positions of the β and β' subunits of RNA polymerase and wild-type β-galactosidase (not present in this experiment). The variation in the relative amounts of fusion protein probably reflects the ratio of parental (Lac⁻) to mutant (Lac⁺) plasmid since no effort was made to segregate the Lac⁺ plasmids. The Lac⁺ mutants were isolated by H.-S. Shin, M. Berberich, P. Wolfe, B. Brownstein, D. Boxer, F. Warren, R. Levesque, W. Marcotte, D. Eichinger, M.-E. Armengod, R. Saunders, and E. Yakobson. (Gel provided by S. Weisbrod.)

EXPERIMENT 3
Cloning *lacZ* Fusions on a High-copy-number Plasmid

In Experiment 1 we obtained λ transducing phages that carry the various *ara-lac* fusions. If we take the size of the *lacZ* gene as approximately 3 kb, this represents about 6% of the λ transducing phage DNA. It should be clear that, by subcloning the fusion and reducing the ratio of vector DNA to insert DNA, we can facilitate physical mapping, DNA sequence analysis, and most in vitro techniques. The purpose of this experiment is to demonstrate a simple method for subcloning any *lacZ* fusion.

A widely used plasmid with a high copy number is pBR322 (Bolivar et al. 1977; see Appendix I, p. 244). This vector carries genes that code for resistance to ampicillin and tetracycline. Since the complete DNA sequence of this plasmid is known (Sutcliffe 1979; Peden 1983), it is possible to predict the pattern of restriction fragments obtained with various enzymes. The copy number of the plasmid is also relatively high (~20 copies/chromosome), therefore cloning of genes in pBR322 often leads to amplification of the cloned gene products. In addition, use of this vector facilitates purification of large amounts of DNA for analysis. Finally, if the *lacZ* gene is cloned in pBR322 (4362 bp), the bacterial insert would represent approximately 40% of the total DNA. These properties make pBR322 ideal for most genetic engineering projects.

There is a small pBR322 derivative—pMLB524 (see Appendix I)—specifically designed for subcloning *lacZ* fusions from λ transducing phages (Berman et al. 1984). In this plasmid the tetracycline-resistance determinants, which lie between the *Eco*RI and *Ava*I sites of pBR322, have been replaced with a 309-bp fragment from the *lac* operon. This fragment was subcloned, taking advantage of the naturally occurring *Eco*RI site in *lacZ* (codon 1006) and the *Ava*I site in *lacY* (codon 70) (see Appendix M, p. 266).

Plasmid pMLB524 is LacZ⁻ since the fragment from *lacZ* encodes only the 17 carboxyterminal amino acids of the structural gene. The naturally occurring *Eco*RI site at codon 1006 of *lacZ* is the only *Eco*RI site in *lacZ* and in pMLB524. DNA fragments that carry the first 1006 codons of *lacZ* and terminate at this *Eco*RI site do not encode an active β-galactosidase. In fact, removal by nonsense mutation of as few as 10 amino acids from the carboxyl terminus of LacZ results in the synthesis of an inactive monomer. However, if *Eco*RI fragments from the *lac* region are subcloned

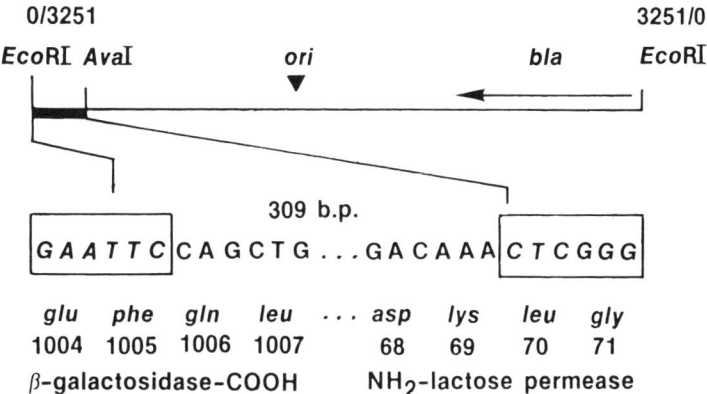

Figure 6
Structure of pMLB524, a fusion cloning vector derived from pBR322. The sequences from EcoRI (position 4361) to AvaI (position 1425) of pBR322 have been replaced with a fragment from the lac operon. The details of this 309-bp lac fragment are shown. The numbers refer to the amino acid residue in β-galactosidase and lactose permease. (ori) pBR322 origin; (bla) β-lactamase gene.

into pMLB524, the plasmid vector will provide the DNA sequence for a proper carboxyl terminus and thereby restore LacZ activity (Fig. 6).

If we consider the locations of the EcoRI sites in the DNA of λ transducing phages, it is possible to predict three general patterns of lacZ fragments (Fig. 7) characteristic of three general classes of fusion transducing phages. In the first type (class I), no additional EcoRI sites are introduced by bacterial DNA substitution. The second and third types of transducing phages would contain additional EcoRI sites located within the bacterial DNA sequences. In class II, the EcoRI site is adjacent to, but not within the target gene. Class III is similar to class II except that there would be an EcoRI site located within the target gene.

Digestion of the DNA from a fusion transducing phage with EcoRI will result in a unique restriction fragment carrying the fusion joint. One end of this fragment is at the EcoRI site in lacZ, the other is in bacterial (class II or III) or phage (class I) DNA. For class I or II, this fragment will carry the regulatory region and the 5' end of the target gene. As mentioned above, such fragments alone do not encode functional β-galactosidase and therefore, if cloned, would not make the cell LacZ$^+$. However, insertion of such fragments in the proper orientation into the EcoRI site of pMLB524 reconstructs an entire lacZ gene, and the resulting plasmid will specify a LacZ$^+$ phenotype.

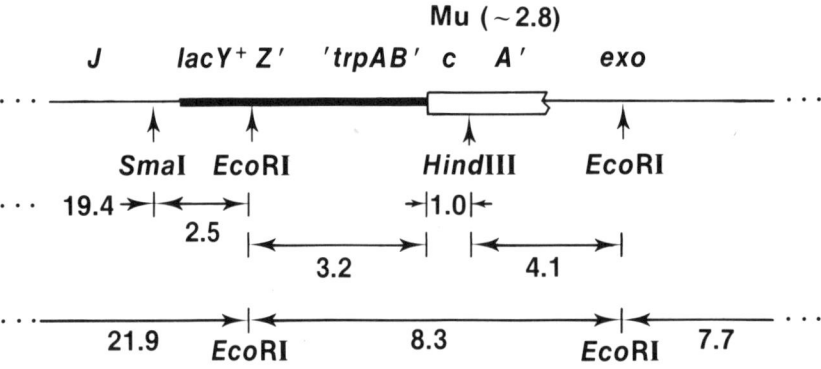

Figure 7
(*Top*) Structure of λpl(209) and related fusion transducing phages. The method of Casdaban (1976a) utilizes the phage λp1(209) to translocate *lacZ* to a target site for the construction of gene fusions. Since our transducing phages are all derived from this phage, certain *Eco*RI sites (↑) are conserved. The two *Eco*RI sites in the right-hand arm of the transducing phages produce fragments identical with λpl(209). These are the sites at 81% (*O*) and 93.1% (*S*). The *Eco*RI site at 65.6% (*exo*) is also generally conserved but may be replaced in certain transducing phages. The two *Eco*RI sites in the *b2* region of wild-type λ have been replaced by the *lac* DNA in both λpl(209) and the fusion transducing phage. The *Eco*RI site within *lacZ* results in an *Eco*RI fragment from the left arm of 21.9 kb, and a fragment carrying most of *lacZ*, the fusion joint, and at least some of the gene to which *lac* is fused. In λpl(209) this is *trp*, designated by an open box. In the transducing phage, the bacterial DNA (3–10 kb) (– – – –) may or may not contain additional *Eco*RI sites (see text). (▨) DNA from the *c* end of Mu (see Fig. 1), which may be present in certain fusion transducing phages, as well. A map of restriction enzyme sites on λpl(209) was initially published by Leathers et al. (1979). (*Bottom*) Map of the restriction enzyme recognition sites in the *lac* region of λpl(209). Sizes of indicated fragments are given in kilobases.

Class III transducing phages present a special problem. If the transducing phage carries an operon fusion, the appropriate EcoRI fragment will always generate an intact *lacZ* gene. However, the promoter region of the target gene will not be present. Accordingly, expression of β-galactosidase will be low. Nevertheless, because of the high plasmid copy number, this expression can be detected using Xgal. If the transducing phage carries a protein fusion, the *lacZ* EcoRI fragment will not generate an intact hybrid gene. In this last case, it is possible to utilize partial digestion by EcoRI or other cloning vectors similar to pMLB524 but that employ different unique restriction sites in *lacZ* (Berman et al. 1984).

The specific goal of this experiment is to isolate DNA from an *ara-lac* fusion transducing phage (see Experiment 1) and subclone the fusion into plasmid pMLB524. Plasmids will be selected as Ap^r, LacZ$^+$ transformants, using the indicator dye Xgal. In this experiment, equimolar amounts of EcoRI-digested phage DNA and plasmid DNA should yield about 50% LacZ$^+$ transformants. These plasmids will be analyzed by restriction enzyme digestion to identify the corresponding EcoRI fragment from the transducing phage. An analytical agarose gel of a typical LacZ$^+$ clone from a fusion phage is shown in Figure 8.

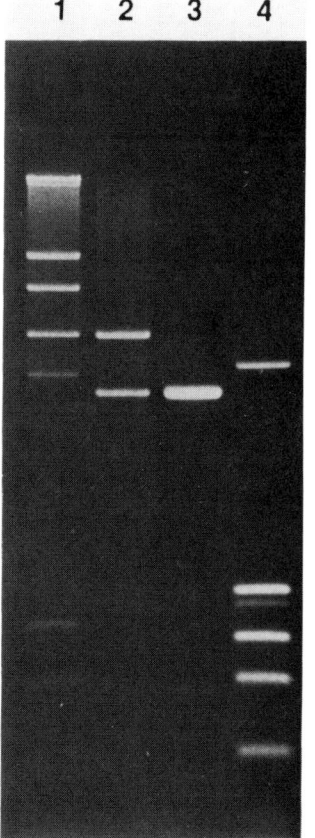

Figure 8

Agarose gel electrophoresis of EcoRI-digested DNA from clones of a gene fusion. This is a photograph of a 0.8% agarose gel stained with ethidium bromide and visualized by UV light. (Lane *1*) An EcoRI digest of a λ fusion transducing phage. The sizes (in kb) of the fragments are: 21.9, 7.4, 5.8, 4.3, 3.5, 1.1, and 0.9. The bacterial DNA carried by this phage must contain two EcoRI sites in addition to the one present in *lacZ*. (Lane *2*) An EcoRI digest of a Lac$^+$ clone derived from this phage into pMLB524. This digest demonstrates that the 4.3-kb fragment from the transducing phage carries most of *lacZ* and the fusion joint. (Lane *3*) An EcoRI digest of pMLB524 (3.3 kb). (Lane *4*) Size standards (in kb): 3.73, 1.35, 1.26, 1.08, 0.87, 0.60, 0.39, and 0.31. (Gel provided by M. Hall.)

Experiment 3

Specific Comments

Strain MC1000 is deleted for *araC*. Accordingly, the level of β-galactosidase expression from a cloned *araBA-lac* fusion may be quite low. This is not a serious problem since the indicator Xgal is quite sensitive. If you clone an *araBA-lac* fusion into strain MC1000 and β-galactosidase activity is inducible by arabinose, you must have cloned *araC*, as well. What does this suggest about gene order, the direction of transcription of *araBAD*, and the distribution of *Eco*RI sites in the region?

DAY 1

Grow a 5-ml liquid lysate of the Lac$^+$ transducing phage (see Procedure 1, p. 89). Prepare phage DNA (see Procedure 27, p. 142) and digest with *Eco*RI (see Procedure 43, p. 183). Use wild-type λ DNA as a control. Digest an aliquot of pMLB524 DNA with *Eco*RI. Run a sample of the digests on an agarose gel to determine if they are complete. Mix the digested phage DNA and a portion of the vector DNA. The amount of vector DNA should be equal to the amount of transducing phage DNA. Extract with phenol/chloroform, and precipitate with ethanol (see Procedures 40, 41, pp. 177, 180). Incubate with DNA ligase at 16°C overnight (see Procedure 31, p. 154).

Inoculate a 5-ml culture of strain MC1000.

DAY 2

Subculture MC1000 and prepare competent cells (see Procedure 37, p. 169). Transform MC1000 cells with 10 µl of the ligated DNA. Select transformants on L plates containing ampicillin (125 µg/ml) spread with 0.1 ml of Xgal.

DAY 3

Purify blue (LacZ$^+$) Ap^r transformants. Check for arabinose inducibility by restreaking on L plates containing Xgal with and without arabinose (add 0.1 ml of 20% arabinose stock per plate). Prepare plasmid DNA (see Procedure 29A, p. 147) and digest with *Eco*RI. Analyze DNA by agarose gel electrophoresis (see Procedure 43, p. 183).

EXPERIMENT 4
Analyzing Chromosome Structure by Using Gene Fusions and DNA Hybridization

The general purpose of this experiment is to locate restriction enzyme cleavage sites within a precise region of a bacterial genome with the aid of gene fusions. Once a *lac* fusion to a specific locus has been obtained, it can be used to isolate and purify DNA from the target gene or DNA corresponding to a regulatory gene. The fusions isolated as described in Experiment 1 contain a λ prophage integrated in the chromosome at a position adjacent to the fusion. Induction of this prophage will yield transducing phages that carry the fusion and often DNA located on either side of the prophage, as well (see Experiments 1, 14, pp. 7, 83). Alternatively, the Lac phenotype conferred by the fusion can be exploited to obtain regulatory mutants (Bassford et al. 1978; Weinstock et al. 1983), and these mutants in turn can be used as indicator strains for screening a library of cloned chromosomal DNA for a λ transducing phage carrying a regulatory gene (see Experiments 5, 7, pp. 39, 47). DNA from either the fusion transducing phage or the transducing phage constructed in vitro can be labeled directly, or the region of interest can be subcloned on a plasmid prior to labeling (see Experiment 3, p. 28, and see below). This labeled DNA can then be used as a radioactive probe in hybridization experiments to analyze bacterial DNA digested with a variety of restriction enzymes. The fragments that hybridize to the probe can be used to construct a map of restriction enzyme sites at the target locus.

Knowing the presence (or absence) and size of DNA restriction fragments homologous to a specific probe not only facilitates the construction of a map of restriction sites, but also makes predictions that can be used for subsequent cloning. For example, the techniques described in this experiment can be used to determine the most appropriate restriction enzyme and vector system to assure that a particular region is represented in a recombinant DNA library. These techniques can be used to examine genes in other genera and species of bacteria. DNA homologs can be isolated on phages (see Experiment 6, p. 43) and plasmids and tested for function (see Experiment 7). Such related loci often give insight into gene structure and function that cannot be obtained by analysis within a given species. Finally, DNA hybridization can be used to determine chromosome structure in mutants containing insertions (see Experiment 8, p. 53), deletions (see Experiments 10, 11, pp. 63, 71), or inversions.

The specific goals of this experiment are to construct a map of restriction sites at the *ompB* locus in *Escherichia coli* (see Appendix N, p. 283) and to compare the structure of the *ompB* locus in *E. coli* with any homologous regions found in several other gram-negative bacteria. The radioactive probe used in this experiment will be a DNA fragment containing the *E. coli ompB* locus (Fig. 9).

The *ompB* locus specifies the regulatory elements that control expression of the porins, OmpF and OmpC. The DNA to be used as a probe in this experiment was obtained in two steps that rely totally on gene fusions. First, fusions to the structual genes *ompF* and *ompC* were isolated (Hall and Silhavy 1979, 1981a; see Experiment 1). Then, the Lac phenotype of these fusions is used to screen a library of λ transducing phages made by in vitro recombination (see Experiment 5) for those phages carrying the regulatory locus, *ompB* (see Experiment 7). One of these transducing phages is the source of the *ompB* probe (Taylor et al. 1981). As mentioned above, it is also possible to use a λ fusion transducing phage as a source of a specific DNA probe (see Experiment 1). For an example of how this was successfully applied to *ompC*, see Mizuno et al. (1983).

The steps involved in this experiment can be summarized as follows:

1. Construct a fusion.
2. Clone a specific DNA fragment defined by the fusion as probe.
3. Digest the bacterial DNA of interest with a set of restriction enzymes. Separate these fragments by agarose gel electrophoresis. Hybridize the radiolabeled probe to these fragments.
4. Construct a map of those fragments that show homology to the probe.

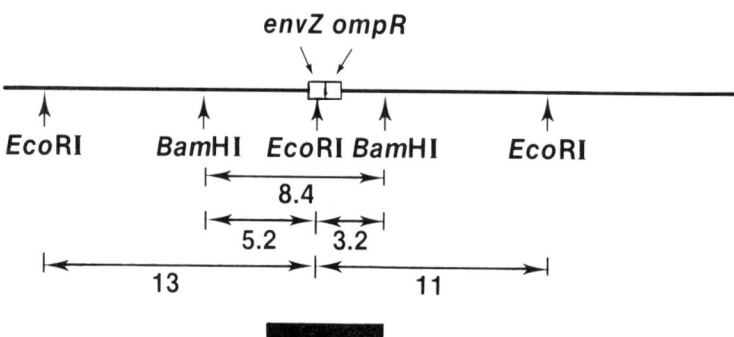

Figure 9
A map of the restriction enzyme recognition sites in the *ompB* locus of *E. coli*. Sizes of the indicated fragments are in kilobases. (☐) The *envZ* and *ompR* genes. (■) A 5.5-kb *Bam*HI-*Sal*I fragment cloned in pBR327 (pRT516); this plasmid is used as a radioactive probe in Experiments 4 and 6 (pp. 33,43).

In this experiment, DNA will be isolated from *E. coli* and also from several strains of gram-negative bacteria, including *Salmonella typhimurium*, *Proteus mirabilis*, *Klebsiella pneumoniae*, *Serratia marcesens*, *Enterobacter cloacae*, and *Shigella sp.* Aliquots of the DNA will be digested with each of three restriction enzymes: *Eco*RI, *Bam*HI, and *Hinc*II. The digested DNA will be fractionated by size, using agarose gel electrophoresis. A DNA fragment from the *ompB* locus (Fig. 9) will be labeled with ^{32}P and hybridized with the DNA fragments in the agarose gel. We will demonstrate two procedures to accomplish this: the Southern transfer method,[1] and the gel dry-down method. The former is a now-classic technique for the transfer of denatured DNA fragments from an agarose gel to a nitrocellulose filter. The filter is dried and hybridized with denatured, radioactive probe. After washing to remove nonspecific hybridization, the filter is exposed to an X-ray film, and areas of hybridization are revealed as dark bands on the developed film. The latter technique relies on the drying of a gel containing denatured DNA fragments. Hybridization of the denatured, radioactive probe is done directly with the dehydrated gel. The hybridized gel is then washed as usual and placed on X-ray film exactly as in the Southern transfer method. In both methods, after examination of the X-ray film results, one can determine the number and size of fragments generated by each restriction enzyme that are homologous to the *ompB* probe (see Fig. 10).

---

[1]The Southern transfer method is an adaptation of the technique first described by E. Southern (1975). For a more detailed description of procedures for restriction enzyme digests, blotting, hybridization, and general nucleic acid chemistry, see Wu (1979), Maniatis et al. (1982), and also M. Barinaga et al., 1981, "Methods for the transfer of DNA, RNA and protein to nitrocellulose and diazotized paper solid supports," Schleicher & Schuell, Inc.

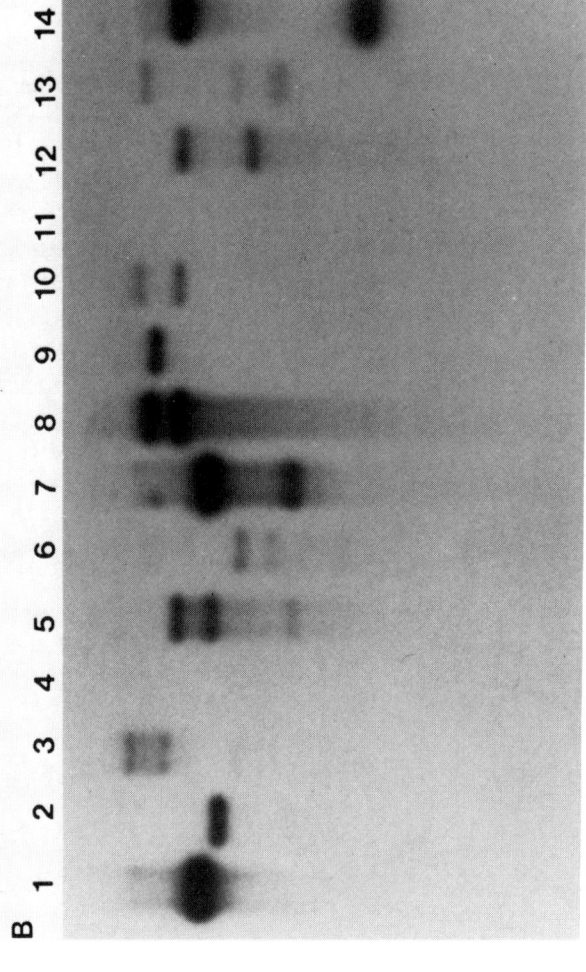
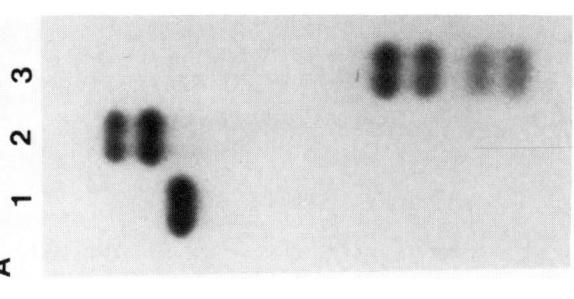

Figure 10

(A) Southern blot of *E. coli* DNA digested with various restriction enzymes and hybridized with a probe from the *ompB* locus. This photograph of an X-ray film shows the various restriction fragments homologous to the *ompB* probe. Digests: (Lane 1) *Bam*HI; (lane 2) *Eco*RI; (lane 3) *Hinc*II. These data, plus data obtained from an *Eco*RI-*Bam*HI double digest, were used to construct the map in Fig. 9. (Data obtained by H.-S. Shin, M. Berberich, D. Boxer, and F. Warren.) (B) Southern blot of DNA from various bacteria digested with either *Bam*HI or *Eco*RI and hybridized with a probe from the *ompB* locus. Digests: (Lane 1) *E. coli*, *Bam*HI; (lane 2) *E. coli*, *Eco*RI; (lane 3) *E. cloacae*, *Bam*HI; (lane 4) *E. cloacae*, *Eco*RI; (lane 5) *K. pneumoniae*, *Bam*HI; (lane 6) *K. pneumoniae*, *Eco*RI; (lane 7) *P. mirabilis*, *Bam*HI; (lane 8) *P. mirabilis*, *Eco*RI; (lane 9) *S. typhimurium*, *Bam*HI; (lane 10) *S. typhimurium*, *Eco*RI; (lane 11) *S. marcesens*, *Bam*HI; (lane 12) *S. marcesens*, *Eco*RI; (lane 13) *Shigella*, *Bam*HI; (lane 14) *Shigella sp.*, *Eco*RI. (Gel provided by S. Bear.)

Specific Comments

It is important to ensure that the bacterial DNA is completely dissolved before restriction enzyme analysis.

Students may see that fragments smaller than 1000 bp tend to be lost in the gel dry-down method. This is not a problem with the Southern blot.

Map construction relies on the hybridization of overlapping fragments, a requirement that often takes several experiments to attain. In addition, the usefulness of any hybridization technique is only as good as the specificity of the probe. Some students forget that the radioactive probe DNA often contains more than just the gene of interest. This mistake yields some interesting conclusions and provides another opportunity to stress the value of controls.

It should be obvious that one could use *lac* DNA as a radioactive probe to analyze the physical structure of a locus defined by gene fusions. How would you design such an experiment?

DAY 1

Prepare a ^{32}P-labeled *ompB* probe by nick translation (see Procedure 48, p. 199). Begin DNA isolation from *E. coli* and the bacterium of your choice (this step corresponds to Experiment 5, Day 1, p. 42). Follow Procedure 25 (p. 137). Make sure the bacterial DNA is completely dissolved before proceeding.

DAY 2

Estimate the concentration of your DNA preparations on agarose gels, using a known amount of DNA as a standard (see Procedure 25). Digest bacterial DNA samples with restriction enzymes *Eco*RI, *Bam*HI, and *Hin*cII. Verify that the restriction enzymes worked and then run your digested samples on agarose gels overnight (see Procedure 43, p. 183).

DAY 3

Stain the gels with ethidium bromide and photograph (see Procedure 43). Use one gel for the Southern transfer method (see Procedure 44, p. 186). Use the other gel for the gel dry-down method (see Procedure 45, p. 189). After this gel has dried, prehybridize it and then hybridize with the radioactive *ompB* probe overnight (see Procedure 46, p. 191).

DAY 4

Continue with the Southern transfer method and bake the nitrocellulose filter blot (see Procedure 44). Prehybridize the filter and hybridize with the radioactive *ompB* probe overnight (see Procedure 46). Continue with the gel dry-down procedure and wash the gel. Expose the gel to X-ray film overnight (see Procedure 46).

DAY 5

Continue with the Southern transfer method and wash the hybridized filter. Expose the filter to X-ray film overnight. Develop the X-ray film from the gel dry-down method.

DAY 6

Develop the X-ray film from the Southern transfer method. Examine the exposed films and, if necessary, reexpose either the filter or dried gel.

DAYS 6-8

Compare your hybridization results with those shown in Figure 10. Examine the exposed films and calculate the apparent molecular weights of the hybridized fragments, using appropriate molecular-weight standards. Compare the calculated fragment sizes with the known map of the *ompB* locus (Fig. 9).

EXPERIMENT 5
Constructing λ Transducing Phages by Using Recombinant DNA Techniques

Specialized transducing phages have been a mainstay of bacterial genetics. They are valuable tools for moving genetic markers (see Appendix J, p. 253), for obtaining defined mutations (see Experiments 11, 12, pp. 71, 75), for genetic mapping (see Experiments 13, 14, pp. 79, 83), and for isolating and purifying DNA for biochemical analysis (see Experiments 3, 4, pp. 28, 33). Nearly all specialized transducing phages for *E. coli* genetics are lambdoid phages, and most of these are derivatives of the bacteriophage λ. Using *E. coli* as host, as much as 40% of the λ genome can be replaced by heterologous DNA without affecting phage growth (see Appendixes F, I, pp. 236, 244). This somewhat surprising fact has allowed the isolation of numerous plaque-forming λ phages carrying a variety of *E. coli* genes.

Until recently, methods for isolating specialized λ transducing phages required the existence of an attachment site (see Appendix F), either primary or secondary, adjacent to the gene of interest. Integration of the phage at this site, followed by an aberrant excision event, could yield a specialized transducing phage (Shimada et al. 1975). Such sites, although numerous, are limited. The discovery of transposable elements overcomes this limitation in *E. coli* since these elements can provide homology or a mechanism for λ integration at random sites located throughout the chromosome (see Experiment 1, p. 7; Appendixes J, L, pp. 253, 261). Moreover, with the advent of recombinant DNA technology, specialized λ transducing phages carrying DNA from virtually any organism can now be constructed. In this experiment, the latter methodology will be illustrated.

The specific goal of this experiment is the construction of transducing phages in vitro, using a specific cloning vector called λD69 (Mizusawa and Ward 1982; see Appendix I). We will insert DNA fragments, generated by the restriction enzyme *Bam*HI, from a variety of gram-negative bacteria into this vector. The recombinant molecules will be packaged into phage coats in vitro, producing infectious phage particles (Sternberg et al. 1977). This collection of transducing phages, most of which carry a single, unique *Bam*HI fragment, will comprise a library or clone bank for the particular bacterial species employed. In following experiments, we use these libraries to isolate transducing phages that carry a particular gene of interest (see Experiments 6, 7, pp. 43, 47).

The versatile λ vector chosen for the experiment, λD69, is designed for insertion of *Bam*HI or *Hin*dIII fragments (see Appendix I). The vector is particularly useful for *E. coli* genetics because it contains the site-specific recombination system of λ. There is a single *Bam*HI site within the *int* gene. Insertion of a DNA fragment at this site inactivates the *int* gene, an event that can be recognized by a plaque color test (Enquist and Weisberg 1976; see Procedure 5, p. 97, and Plate 3). These phages no longer can integrate unless active Int is supplied by a helper phage. Once these Int⁻ phage are integrated by site-specific recombination, they are "locked in" (see Appendix H, p. 241), an event that has useful genetic consequences. Because λD69 *Bam*HI hybrids are defective for Int, in the absence of helper phages they can only form stable lysogens by homologous recombination. Again, this is an event that has useful genetic consequences (see Appendix J).

Students should have prepared bacterial DNA samples and should know the approximate concentrations (see Experiment 4). Each bacterial DNA sample, as well as the λD69 DNA, is digested with *Bam*HI. The digested DNA samples and vector will then be ligated. Two important controls for these ligation reactions are digested vector alone with and without DNA ligase. Each ligation reaction is tested by agarose gel electrophoresis, packaged in vitro, and amplified by preparation of a plate stock. The resulting recombinant phages are screened using the red-plaque test. In subsequent experiments, we screen these libraries for specific transducing phages (see Experiments 6, 7).

Specific Comments

The library made in this experiment represents a large percentage of the bacterial genome and indeed is adequate for most experimental purposes. However, students should realize that not all restriction fragments are present in the library. Some fragments are too small or too big for the vector (see Appendix I, p. 244). A more representative library is obtained when large, overlapping DNA fragments are inserted in the λD69 vector. This method is described in Procedure 34 (p. 162) and should be used if a truly representative library is needed.

It is possible to calculate the number of individual clones needed in a library to represent any given gene with a 99% probability (Carbon et al. 1977; Seed 1982; Seed et al. 1982). This is calculated according to the following formula:

$$P = 1 - (1 - F)^N \quad \text{or}$$

$$N = \frac{\ln(1 - P)}{\ln(1 - F)}$$

Where:

P = probability of any unique sequence being present in the library;
N = number of hybrids in the library;
F = fraction of the total genome in each hybrid; e.g., the size of the average cloned fragment divided by the total genome size.

Assume a bacterial genome has a size of 5000 kb and each λ hybrid carries a DNA fragment of 10 kb. Then a complete library (e.g., $P = 0.99$) would be contained in no more than 2500 independent phages.

$$N = \frac{\ln(1 - 0.99)}{\ln(1 - 10/5000)}$$

$$N = \frac{\ln(0.01)}{\ln(0.998)}$$

$$N = 2302$$

The initial number of plaque-forming phages in the in vitro packaging extract is small, usually 10^4-10^6. By plating about 10^4 phages and making a plate stock, you should obtain a lysate of about 10^{10}/ml. This amplified library should be a source of transducing phages for a long time. Some cloned fragments affect λ growth, in which case these clones could be under- or overrepresented in the amplified library.

Students occasionally use liquid lysates rather than plate stocks to amplify their primary, in vitro packaging lysates. This is a good way to demonstrate natural selection and survival of those that produce the most progeny. Plate stocks made from plates in which individual plaques do not overlap will contain representatives of each plaque-forming particle in the lysate, no matter how poorly or how well they grow. Accordingly, this is the method of choice. Successive plate stocks, however, lead you down the same path as liquid lysates.

Strains like LE392 are essential for library construction when the DNA to be cloned in either phages or plasmids is derived from organisms other than *E. coli* K-12. LE392 is restriction-minus, modification-plus for the K restriction system (*hsd*). This means that DNA lacking the K modification will not be degraded in LE392 but, upon replication, will be K-modified. If the in vitro packaging lysates are plated directly on *E. coli* K-12 strains that are restriction-plus, the unmodified DNA will be degraded and few phages will survive.

Using the methods described here, one should expect to make packaging extracts with efficiencies of 10^7-10^8 plaques/μg of wild-type λ DNA. Amplified libraries routinely contain 50–80% hybrid phages, as determined by the red-plaque test.

Packaging extracts often yield low levels of small, clear plaques (10^2-10^3 plaques/ml). These presumably represent aberrant excision events of the defective prophages in the packaging strains.

42   Experiment 5

DAY 1

Isolate bacterial DNA (see Procedure 25, p. 137; this is also the same as Experiment 4, Day 1, p. 37). Make sure the bacterial DNA is completely in solution before proceeding.

DAY 2

Digest bacterial DNA and vector DNA with *Bam*HI (see Procedure 30, p. 152). Verify that digestion is complete by running a rapid agarose gel (see Procedure 43, p. 183). If digestion is complete, set up ligations for overnight incubation (see Procedure 30). Prepare a 5-ml overnight culture of LE392 in TB containing 0.2% maltose.

DAY 3

Run samples of each ligation reaction on an agarose gel to check ligation (see Procedure 43). Centrifuge LE392, resuspend in 0.5 volume of 10 mM $MgSO_4$, and store at 4°C. Package phage DNA from ligation reaction (see Procedure 39, p. 173). Titer in vitro packaging reactions using LE392. Prepare a 5-ml overnight culture of LE292 in TB for the red-plaque test.

DAY 4

Examine the plates and calculate the titer of each primary, in vitro packaging reaction. Make plate stocks using LE392 (see Procedure 2, p. 91). Determine the percentage of hybrid λD69 phage, using the red-plaque test, and strain LE292 (see Procedure 5, p. 97).

DAY 5

Determine plate stock titers and calculate the number of hybrids.

EXPERIMENT 6
Identifying λ Transducing Phages by Using DNA Hybridization

DNA hybridization is an effective technique for identifying a particular DNA sequence. Experiment 4 (p. 33) describes ways of generating unique DNA probes from different chromosomal regions by using gene fusions. These probes are used to determine the chromosomal structure at the *ompB* locus. Employing one of these techniques to generate a suitable DNA restriction fragment, one can label and use the fragment to identify clones, either in plasmid or phage vectors, that share sequence homology. For example, a probe from the 5' portion of a gene fusion can be used to locate clones from a library (see Experiment 5, p. 39) that carry the intact target gene. In contrast to procedures using genetic complementation to identify clones (see Experiment 7, p. 47), the type of hybridization protocol described here can be implemented when the target gene exhibits no easily selectable or screenable phenotypes. It can also be used to identify clones of homologous gene sequences (at the DNA level) that do not necessarily function in the same manner as the target gene. Such experiments permit the isolation of families of DNA sequences from numerous organisms without demanding functional homology. This may provide information about relatedness and evolution of function. Finally, clones that carry only a portion of the probed gene can be detected.

There are two general methods for examining clones by hybridization. The first procedure involves the transfer to a nitrocellulose membrane of bacterial colonies transformed by a library of plasmid clones. Appropriate clones are detected by subsequent lysis of the cells and hybridization with radiolabeled DNA probes (Grunstein and Hogness 1975). To identify particular members from a library of λ clones (see Experiment 5), a slightly different method is employed. In this case phage plaques representing the total library are allowed to develop in a bacterial lawn. Subsequent transfer to a membrane support and hybridization with a probe will reveal which plaques (i.e., which clones) carry a gene sequence of interest (Benton and Davis 1977). With both methods, it is possible to obtain viable clones (phages or colonies), following the hybridization step, by using the "positive" radioactive signals as a guide. In most experiments the first screening must be confirmed by a second round of hybridizations. In this way, false positive clones can be eliminated.

This experiment uses the method described by Benton and Davis (1977). The success of this protocol relies on certain features of λ physiology. When λ packages its DNA in the infected cell, it is less than 100% efficient. In fact, for wild-type λ, about 25% of the λ DNA in the cell is not packaged. This number increases to about 50% if the λ is defective in the *exo* or *gam* gene (Enquist and Skalka 1973). The unpackaged DNA is released upon lysis and remains within the plaque. This free λ DNA as well as phage particles can be transferred to nitrocellulose paper by simply dropping a filter directly onto a lawn of plaques (Benton and Davis 1977). The filter is gently lifted off, leaving an intact lawn with still viable plaques and a replica of the plaques on the filter. After denaturation and fixation of the plaque DNA to the nitrocellulose filter, hybridization to the DNA probe of interest is performed. After hybridization, the "blots" are autoradiographed and positive regions are aligned with the plaques. Putative positive plaques are picked from the original plate, reblotted, and hybridized for verification. Pure positive phages are isolated and plate stocks are made.

There is one important fact to keep in mind. The host cell, in which the plasmids are grown or on which the phages are plaqued, must *not* have extensive homology with the experimental probe. With gene fusions, it is possible to isolate a probe (see Experiment 4) as well as a chromosomal deletion of the probe sequences from the genome (see Experiment 10, p. 63). It is also possible to construct in vitro an internal deletion of target gene sequences from clones of gene fusions. Such a deletion can be crossed onto the chromosome by homologous recombination (see Appendix J, p. 253) if there is homology with the target gene on both sides of the deleted region. If such a deletion is made by removing a defined fragment of DNA, then this fragment can be subcloned and used as a hybridization probe in the deletion strain. Such chromosomal deletions will reduce the background hybridization (or noise) that may obscure any positive signals.

The specific goal of this experiment is to examine a mixture of λ phage plaques by DNA hybridization. We will try to detect phages that carry sequence homologies with the *ompB* locus (see Appendix N, p. 283). The probe used is specific for the *ompB* region; consequently the host cell carries a deletion in this region of the chromosome. This is the same probe used in Experiment 4.

Specific Comments

Although the general approach as outlined above is designed for screening a library of phage clones, we have designed this experiment to demonstrate the protocol with an artificial mixture of λ transducing phages. The reasons for this are twofold. First, the method of choice for obtaining an OmpR⁺ clone is selection by complementation (see Experiment 7, p. 47). As always, a selection is preferable to a screen of candidates. Second, to minimize the use of radioactivity, we will not rehybridize the positive plaques but instead will rely on screening another phage phenotype. An appropriate mixture contains the following two phages. One phage, λTK10, will hybridize with the *ompB* probe; in addition, this phage is also LacZ⁺. Therefore, plaques isolated as positive signals by using the labeled probe can be verified by using Xgal indicator plates. The other phage in the mixture, λ wild type, will not hybridize with the probe and, of course, is LacZ⁻. An example of a plaque blot obtained previously is shown in Figure 11.

If time permits, students should be encouraged to use the plaque-blot procedure and the OmpB probe to examine the libraries constructed in Experiment 5 (p. 39).

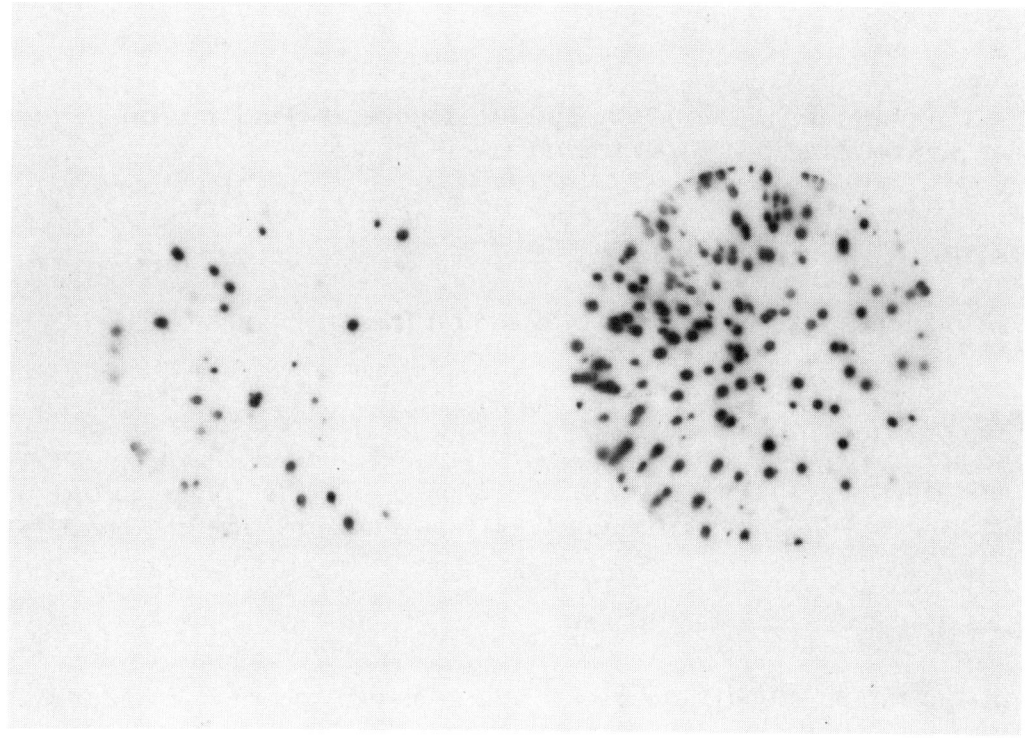

Figure 11
X-ray film showing two plates that exhibit positive plaque hybridization. It was obtained by J. Ruether and T. Fekete, using the techniques described in this experiment, with a ratio of λ wild type to λTK10 of ~ 2:1. The results of two different dilutions of the phage lysate are shown.

46 Experiment 6

DAY 1

Read Procedure 47 (p. 196) and inoculate a 5-ml culture of SG480 late in the day. This strain, like other *malQ* strains, is somewhat maltose-sensitive. Do not add maltose to the growth media. Nick translate probe pRT516 (see Procedure 48, p. 199).

DAY 2

Use the phage mix described in Specific Comments. Plate mix on plaque-blot plates, using SG480 as the host.

DAY 3

Blot plaques; bake blots; prehybridize and then hybridize overnight.

DAY 4

Wash filters and autoradiograph. Inoculate 5 ml of overnight culture of MC4100 in TB. Develop the film later in the day or early on Day 5.

DAY 5

Line up the filter and X-ray film. Locate isolated, "positive" plaques and pick with a sterile, pointed toothpick into a top agar lawn of MC4100 cells on L plates. Add Xgal to top agar prior to plating the lawn (see Procedure 8, p. 104). Stab about 50 "positive" plaques and 50 "negative" plaques.

DAY 6

Score results.

EXPERIMENT 7
Identifying λ Transducing Phages by Genetic Complementation

Any gene cloned in a λ vector can be identified by complementation if the gene is expressed and the gene product will confer a recognizable phenotype in *E. coli*. This approach has been used for some time with conventional transducing phages. The combined innovations of recombinant DNA and gene fusion technology bring new power to this now-classic genetic experiment. For example, not only can one isolate genes for which there is no direct selection, but one can also isolate genes from other organisms and test them for function in *E. coli*.

The general purpose of this experiment is to use gene fusions as tools for both selecting and screening recombinant DNA libraries for specific complementing, transducing phages. Several basic genetic techniques are employed, including construction of lysogens by homologous recombination and the subsequent isolation from these lysogens of plaque-forming transducing phages.

The specific purpose of the experiment is twofold: first, to isolate λD69 hybrid phages carrying a DNA fragment encoding the positive regulatory locus, *ompB* (see Appendix N, p. 283), from an *E. coli* library (see Experiment 5, p. 39); second, to isolate λD69 hybrid phages carrying DNA fragments from other enteric bacteria that complement an *E. coli* ompR mutation. The primary phenotype that will be used is Lac. The system is as follows: The *ompR* gene encodes a positive regulatory protein required for expression of the *ompF* and *ompC* genes (see Appendix N). We use a strain in which *lacZ* is fused to the *ompF* promoter. This strain becomes Lac$^+$ when *ompF* is expressed. A nonreverting mutation, *ompR101*, that inactivates the positive regulator has been introduced into this fusion strain (MH5101). Consequently, *ompF* is shut off and the strain is Lac$^-$. If a functional *ompR* gene is introduced into MH5101, the strain will become Lac$^+$. This system provides the means for identifying OmpR$^+$ clones by complementation from the libraries constructed in Experiment 5 (Taylor et al. 1981). We employ two complementation techniques. The first is a selection; i.e., only those phages capable of transducing MH5101 to Lac$^+$ are isolated. The second is a screen; i.e., all the plaque-forming phages in the library are examined on indicator plates where Lac$^+$ and Lac$^-$ plaques can be distinguished.

In the selection procedure, you use lysates of an *E. coli* library as well as of the libraries prepared from other enteric bacteria to lysogenize MH5101. The lysogens are spread on lactose minimal agar. Only those lysogens that are Lac⁺ will grow. Lac⁺ lysogens are purified, and spontaneously released phages are isolated and tested (see Appendix H, p. 241).

In the screening procedure, you again use both the *E. coli* library and enteric libraries. In this case, the phages are plated in a top agar lawn of MH5101 on lactose minimal agar. By permitting only slight growth of MH5101, Lac⁺ complementing phages appear as large, turbid plaques, whereas Lac⁻ phages form small, clear plaques. The slight growth of the lawn is obtained by simply adding a small amount of broth to the minimal top agar.

Although this experiment demonstrates fundamental genetic techniques, it also poses a classic genetic problem: You get what you select. Students isolate phages that confer a Lac⁺ phenotype. The system is set up to identify OmpR⁺ clones, but is our logic infallible? Complementing phages isolated from the libraries, especially the enteric libraries, can yield unexpected results. There are at least eight distinct classes of Lac⁺ "transductants" that may emerge from this experiment, only one of which results from an authentic OmpR⁺ transducing phage. Understanding some classes requires careful thought about the difference between complementation and recombination. The two most obvious possibilities are that you may find (1) a λD69 hybrid carrying *lac* genes, which will certainly complement the Lac⁻ defect of MH5101, or (2) a contaminant. By careful choice of controls, by analysis of the Lac phenotype of spontaneously released phages from the lysogens (see Appendix H), and by analysis of *ompC* expression (another gene positively controlled by the *ompR* gene), you should be able to characterize the various Lac⁺ transductants correctly.

As a further aid we have provided detailed day-by-day protocols that include many important controls. For example, a good positive control is the OmpR⁺ transducing phage λ*p*RT2. An appropriate negative control is the cloning vector λD69.

Specific Comments

In the screening technique, the complementing phage is isolated directly from a plaque, whereas in the lysogen selection technique, the complementing phage is isolated from a lysogen. The latter technique depends on isolation of a stable lysogen, an event that requires integration of the transducing phage into the host chromosome. Since λD69 hybrids are defective in site-specific recombination, lysogens can only be formed by homologous recombination between the phage and the chromosome. If the *E. coli* host contains a λ prophage, recombination within homologous λ segments results in integration. If the *E. coli* host contains DNA homologous to the bacterial sequence carried by the phage, recombination can occur between these segments. In any case, lysogen selection depends at least on two events: recombination as well as restoration of function. Despite this fact, these rare lysogens can still be detected because at least 10^8 to 10^9 infected cells can be plated on selective media.

It is possible to isolate Lac⁺ transductants that are multiple lysogens for OmpR⁺ transducing phages and λD69 hybrids carrying unknown DNA segments. In these strains, the spontaneously released phages will be a mixture. These phages must be purified and retested.

DAY 1 (late in the day)

Inoculate a 5-ml overnight culture of MH5101 in TB at 30°C.

DAY 2

Collect the cells by centrifugation and resuspend the cell pellet in 5 ml of 10 mM $MgSO_4$. Collect the cells again by centrifugation and resuspend the pellet in 2.5 ml of 10 mM $MgSO_4$. Keep the cell suspension on ice or at 4°C.

Lysogen Selection

1. For *each* library, prepare the following in six small, sterile test tubes:

Tube	MH5101	Phage
1	0.1 ml	library lysate, 10^7 phages
2	0.1 ml	library lysate, 10^8 phages
3	0.1 ml	λpRT2, $5 \times 10^3 - 10^4$ phage
4	0.1 ml	λ D69, 10^8 phage
5	0.1 ml	—
6	—	library lysate, 10^8 phages

   Mix well and incubate for 5 min at room temperature for phage adsorption.

2. Add 2.5 ml of molten (45°C) F top agar to all six tubes and pour each on a lactose minimal plate. Allow top agar to solidify and incubate at 37°C for 2 days.

Plaque Screen

1. For *each* library to be screened, prepare seven small, sterile tubes as follows:

Tube	MH5101	Phage
1–5	0.1 ml	library lysate, 500–1000 phages
6	0.1 ml	λpRT2, 500–1000 phage
7	0.1 ml	λ D69, 500–1000 phage

   Add no more than 0.2 ml of lysate dilution per tube and mix well. Incubate at room temperature for 5 min for phage adsorption. Add 0.4 ml of TB (broth, not top agar) to each tube, followed by 2.5 ml of molten (45°C) F top agar.

2. Pour the contents of each tube on a lactose minimal plate and allow it to solidify. Incubate at 37°C for 2 days.

50 Experiment 7

DAY 4

Examine both the lysogen selection and the plaque-screen plates. Be sure that the positive controls give Lac$^+$, as expected, and that the negative controls are Lac$^-$. In the lysogen selection, you plated 5,000–10,000 plaques of λpRT2 as a positive control. You should see only 50–100 Lac$^+$ colonies. Why?

Lysogen Selection

1. Pick eight Lac$^+$ transductants from each library and two from the λpRT2 positive control plates.

2. Streak each transductant for single colonies on lactose minimal plates. Incubate at 37°C for 24 hr.

Plaque Screen

1. Examine each plate for Lac$^+$ plaques. Often the Lac$^+$ cells in a positive plaque cover the entire plaque. There should still be phages, nevertheless. Using a pasteur pipette, pick each Lac$^+$ plaque (do no more than eight) to 0.2 ml of λ-dil containing 0.05 ml of chloroform (this phage suspension is called a pickate). Mix well and let the chloroform settle. (The cells in the center of the Lac$^+$ plaque are lysogens and could be analyzed exactly as described in Lysogen Selection of Day 4.)

2. Streak for single plaques or plate 5 µl and 50 µl of each pickate on lawns of MH5101 on lactose minimal plates. Incubate at 37°C for 24 hr.

3. Inoculate a 5-ml overnight culture of MC4100 in TB containing 0.2% maltose at 30°C.

DAY 5

Examine plates.

Lysogen Selection

Inoculate an overnight culture of each of the eight purified Lac$^+$ transductants in 2 ml of L broth. Incubate at 37°C overnight.

Plaque Screen

1. If the original Lac⁺ plaque arose from an OmpB⁺ transducing phage, then all of the plaques from the initial pickate should transduce Lac⁺. Any "nontransducing" plaques may be the result of contaminating OmpB⁻ phages. If the pickate is not homogeneous, prepare a fresh pickate from a well-isolated Lac⁺ plaque and repeat this test before proceeding.

2. Streak for single plaques or plate 5 μl and 50 μl of each pickate on MC4100 lawns on TB plates with Xgal (see Procedure 8, p. 104). Incubate at 37°C overnight.

DAY 6

Lysogen Selection

1. The Lac⁺ lysogens should only be formed by homologous recombination. Therefore, the supernatants of overnight cultures of the transductants should contain significant numbers of the transducing phages that have been excised by homologous recombination (see Appendix H, p. 241). These phages are recovered as follows: Remove 1 ml of overnight culture to a sterile microfuge tube and add 0.1 ml of chloroform. Mix well. Save the remaining overnight culture at 4°C. Centrifuge the chloroform-treated culture in the microfuge for 1 min. Save the supernatant and discard the pellet.

2. Make a 10^{-2} dilution of each supernatant. Plate 5 μl and 50 μl of both the undiluted and diluted supernatants on lawns of MC4100 on TB plates containing Xgal (see Procedure 8). Incubate at 37°C overnight.

Plaque Screen

1. Examine the Xgal plates. A true OmpR⁺ transducing phage should make white plaques just like λpRT2. λD69 should also make a white plaque. What would a blue plaque indicate?

2. Verify that the isolated and purified plaques transduce MH5101 to Lac⁺ at high frequency. Use lactose minimal plates. Additional proof that you have indeed isolated an OmpR⁺ transducing phage is that it should complement all known *ompR* mutations (see Experiment 13, p. 79).

52   Experiment 7

3. At this stage, you have identified the transducing phage by function. It is also possible to identify it by physical characterization of its DNA. In Experiment 4 (p. 33), you characterized the *ompB* locus by hybridization and mapped some restriction enzyme cleavage sites in the region. By isolating the DNA from the new transducing phage (see Procedure 27, p. 142) and analyzing it by restriction enzyme digestion (see Procedure 43, p. 183), you should be able to correlate the physical map of the chromosomal DNA (see Fig. 9, p. 34) with that of the transducing phage.

4. Make plate stocks (see Procedure 2, p. 91) of the transducing phage.

DAY 7

Lysogen Selection

1. Follow the instructions for the Plaque Screen on Day 6.

EXPERIMENT 8

Isolating Tn*10* Insertions in or near a Gene

Transposons, in particular the drug-resistance elements and certain derivatives of the bacteriophage Mu, have revolutionized bacterial genetics (Kleckner et al. 1977; Appendix C, p. 226). They provide important experimental tools for genetic mapping (see Experiments 13, 14, pp. 79, 83), mutant isolation (see also Experiments 9, 10, pp. 59, 63), and strain construction. In addition, they provide portable regions of genetic homology, and these can be used to direct chromosomal rearrangements (see Experiment 1, p. 7).

The purpose of this experiment is twofold. First, we will isolate a pool of *E. coli* containing drug-resistance elements inserted randomly in the chromosome. Second, we will demonstrate how this pool can be employed to construct strains that contain a drug-resistance element inserted near a particular structural gene, in that gene, or in a gene whose product is necessary for expression of the target structural gene.

The transposable element Tn*10*, which confers tetracycline resistance (Tc^r), will be introduced into *E. coli* by means of a specifically constructed λ phage, λNK561 (Foster et al. 1981). This phage carries Tn*10* as well as mutations that block phage replication (*O*am, *P*am) and integration (*b*221 deletion, *c*I). These mutations prevent the phage from transducing tetracycline resistance to the recipient by any standard means (e.g., lysogeny or plasmid formation). Since the phage shares no homology with the recipient chromosome, the Tn*10* element cannot be inherited by homologous recombination events. Under these conditions, selection is made for tetracycline resistance. The recipients must acquire the Tn*10* element by transposition from the phage to the chromosome. Since the frequency of transposition is low, each Tc^r survivor will carry one copy of the transposon.

A Tn*10* insertion within a gene of interest can be obtained in a single step, as follows. A suitable strain is infected with λNK561, and Tc^r survivors that have lost gene function are selected. Any *E. coli* strain can be employed as a recipient in this experiment. The only qualifications are that it must be λ sensitive and must not be a λ lysogen. If the strain were a λ lysogen, the prophage would provide homology, allowing chromosomal integration of λNK561 by recombination. The lysogen would also contain λ repressor to complement the *c*I mutation in λNK561. This presents a problem when one is working with strains that carry gene fusions. As is known from Experiment 1, nearly all of the gene fusions we employ contain a λ prophage that lies adjacent to the fusion in the chromosome. Accordingly, the λNK561 phage cannot be employed to isolate Tn*10* insertions in most fusion strains.

To overcome this problem, we employ a two-step procedure. In the first step, a strain that is not a λ lysogen is used as the recipient for transposition. A large pool containing approximately 10,000 independent Tc^r colonies is then collected. Subsequently, a P1 lysate is prepared on the pooled population. The result is a generalized transducing lysate capable of transducing any of the Tn*10* insertions present in the pool to any *E. coli* strain. Using this procedure, we can identify Tn*10* insertions that alter expression of gene fusions.

A further advantage of the two-step procedure is that it permits the isolation of Tn*10* insertions that are near any gene of interest even if the insertion does not result in a mutant phenotype. Such insertions can be isolated following a transduction with the P1 lysate prepared on the pool by selecting simultaneously for the gene of interest and tetracycline resistance. Finally, the P1 lysate should be stable for years. If the pool size is large enough, it should contain Tn*10* insertions near every gene and it should contain an insertion within every nonessential gene. Thus, almost any desired Tn*10* insertion can be obtained following P1 transduction by imposing the appropriate selection or screen. In other words, the transposition step need only be done once.

The specific goal of this experiment is to isolate a Tn*10* insertion near or in *ompC* or an insertion within a gene whose product is necessary for *ompC* expression (see Appendix N, p. 283). To do this we employ the strain SG608 as recipient for transduction with the P1 lysate prepared on the Tc^r pooled population. Strain SG608 is deleted for the chromosomal *lac* genes but carries an *ompC-lac* fusion. As such it is OmpC$^-$ and Lac$^+$. Tc^r transductants that simultaneously lose the *ompC-lac* fusion score as Lac$^-$. These can arise from a Tn*10* insertion within *ompC*, in which case they will remain OmpC$^-$ (resistant to phage hy2—hy2^r); or they can arise from a Tn*10* insertion near *ompC*, in which case they must acquire an *ompC$^+$* gene (hy2^s) from the donor. Transductants that lose the *ompC-lac* fusion must also lose the associated λ prophage. Such strains will no longer be immune to λ.

As described in Appendix N, two genes are known whose products are involved in *ompC* expression, *ompR* and *envZ*. Tn*10* insertions in *ompR* have been isolated. We wish to obtain Tn*10* insertions in *envZ*. The problem is, we do not know what phenotype will be conferred by such a mutation. Indeed, it may confer no phenotype at all. Moreover, if *envZ* is an essential gene, an *envZ*::Tn*10* would be lethal. To circumvent this problem, the strain we employ as a recipient, SG608, is an *ompR*, *envZ* diploid. A transducing phage, λ*p*RT2.3, that carries a mutant *ompB* region and the immunity of phage 21 is integrated into the chromosome by homologous recombination with the λ prophage that lies adjacent to the *ompC-lac* fusion (Fig. 12). The *ompR* allele carried by this phage is wild type. However, the *envZ* allele is mutant, i.e., *envZ3*. In a haploid, the

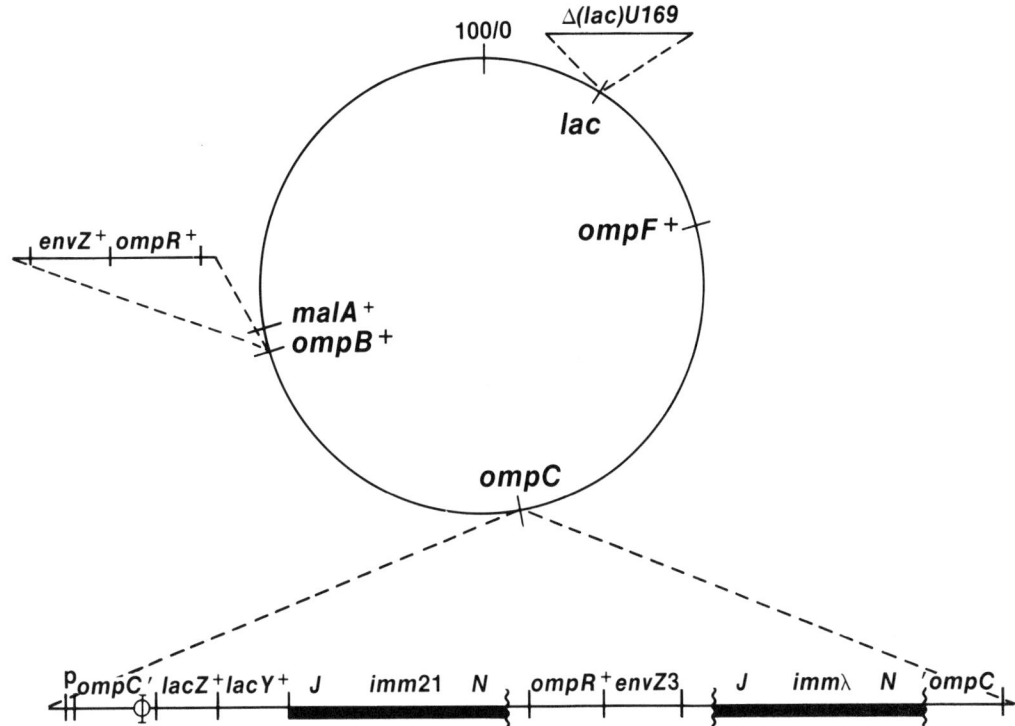

Figure 12
Genetic map of strain SG608, showing chromosomal positions of relevant markers. (▬) λ.

mutant *envZ3* allele confers an OmpC⁻ phenotype.[1] When the mutant allele is present in *trans* to *envZ⁺*, the diploid is OmpC⁺; *envZ3* is recessive.[2] Strains that acquire a chromosomal *envZ*::Tn*10* insertion will lose *envZ⁺* function and thus unmask *envZ3* and the associated OmpC⁻ phenotype. These *envZ*::Tn*10* insertions will be phenotypically Lac⁻ but will still contain the *ompC-lac* fusion. If the fusion is still present, then the associated λ prophage and λpRT2.3 must both be present, as well. Consequently, these strains will remain immune to *imm*λ and *imm*21 phages.

Two final points must be made regarding this experiment. First, since *envZ*::Tn*10* may be lethal, the strain used to prepare a pool of randomly inserted Tn*10* elements must be diploid for *envZ* in order to insure that no essential function is lost. Strain SG404 contains an episome (F′141) that carries *envZ⁺*. In all other respects the strain is essentially wild type.

---

[1] An *ompC-lac* fusion strain, carrying *envZ* is red at 30°C and 37°C on lactose MacConkey agar whereas an *envZ3* mutant appears white at 30°C and pink at 37°C.

[2] Actually, the *envZ3* allele appears recessive only under certain conditions, i.e., at a higher growth temperature, 37°C.

56   Experiment 8

The presence of this episome will not affect transposition in any way, and accordingly, the P1 lysate prepared on the pool can also be used to isolate other Tn*10* insertions.

Finally, students must address the problem of target size. It is much easier to isolate an insertion near a gene than it is to isolate an insertion within a gene. To find *ompC*::Tn*10* or *envZ*::Tn*10*, it will be necessary to score at least 30 of the Lac$^-$, Tc^r strains.

Specific Comments

Partly out of design and partly out of necessity, almost every genetic element and technique described in this manual is used in Experiments 8, 9, and 10. It is important that students understand the logic behind these three experiments. The only assumption made is that a knockout mutation, i.e., an insertion, a nonsense mutation, or a deletion, will be recessive (Austin et al. 1971). Although these experiments use a recessive allele (*envZ3*) to score for a knockout mutation in the wild-type *envZ* gene (Garrett et al. 1983), one should realize that this strategy can be employed with any mutant phenotype whether dominant or recessive. Therefore this strategy can be employed for any gene.

Experience has shown that preparing a P1 lysate on the pooled population of Tc^r transductants can be difficult. This is probably due to the chelating agent (sodium citrate) that is added to prevent λ readsorption. This compound disrupts the *E. coli* cell envelope and interferes with P1 adsorption. To circumvent this problem, we employ a P1Tn*9clr*100 lysogen (strain SG404) as the recipient for transposition. This phage carries Tn*9* and thus confers resistance to chloramphenicol. The *clr*100 mutation is a temperature-sensitive mutation of the P1 repressor. At high temperature, the repressor is inactivated and the phage undergoes lytic growth. By using this strain, a high-titer P1 lysate can be prepared easily by raising the growth temperature. Since this does not require phage adsorption, any problems associated with an altered cell envelope are avoided.

You should note that bacteriophage P1 has its own restriction and modification system. Therefore, if a P1 lysogen is infected with a λ phage with unmodified DNA, it will be recognized as foreign and destroyed. To avoid this, λNK561 should be grown on the strain SG265. This strain is lysogenic for a cryptic (*cry*) P1 phage and as such will modify the λ DNA correctly. Strain SG265 also carries the *supP* allele to allow good suppression of both the *O*am and *P*am mutations.

Note that the lactose tetrazolium–tetracycline agar contains only 8 μg/ml tetracycline. Higher concentrations of the drug in this media inhibit growth of Tc^r strains.

DAY 1

Inoculate a 5-ml overnight culture of SG265.

DAY 2

Plaque purify λNK561 by streaking or plating out dilutions of a lysate onto a lawn of SG265 (see Appendix G, p. 239).

DAY 3

Grow λNK561 from purified plaques (see Procedure 2, p. 91), using strain SG265 as a host. Titer the lysate. Inoculate SG404.

DAY 4

Follow Procedure 17 (p. 119) for transposition of Tn*10* from λNK561 to the *E. coli* chromosome. Use strain SG404 as the recipient. You will notice in this procedure that citrate (20 mM) is present throughout. Citrate is a chelating agent and prevents λ readsorption.

DAY 5

Continue with Procedure 17 and prepare the P1 lysate by heat induction of the pooled population (see Procedure 11, p. 109). Inoculate SG608.

DAY 6

Transduce strain SG608 to Tc^r. This is done as described for P1*vir* (see Procedure 12, p. 111). Since P1Tn*9clr*100 can form lysogens at 30°C, do all transductions at 37°C. Use lactose tetrazolium containing tetracycline (8 μg/ml) agar and incubate at 37°C overnight for selection. Do not forget to allow time to express the Tc^r phenotype.

DAY 7

Score the number of Tc^r transductants. Determine the amount of phage lysate that gives the maximal number of transductants. If the optimum conditions yield more than 200 Tc^r transductants per plate, proceed to scale up the transduction as follows: Prepare 15 separate transductions, using the optimal amount of lysate. Do not forget sterility controls. (The use of multiple transductions is convenient and often yields more total transductants.)

If the lysate yields less than 200 Tc^r transductants per plate, prepare a fresh lysate from the pooled population (see Day 5).

DAY 8

Score results. If you are having trouble identifying Lac⁻ transductants, try replica plating to lactose MacConkey agar containing 8 µg/ml tetracycline at 30°C. Purify at least 30 Tc^r, Lac⁻ transductants on lactose tetrazolium agar containing tetracycline (8 µg/ml).

DAY 9

Repurify Lac⁻ mutants on lactose MacConkey agar at 30°C.

DAY 10

Characterize the Lac⁻, Tc^r mutants as follows:

Site of Tn10 insertion	Prophage immunity	hy2
Near *ompC*	none	s
In *ompC*	none	r
In *envZ*	λ, 21	r

Phage hy2 sensitivity can be tested by cross-streaking on TB agar. For a description of the tests used to demonstrate immunity, see Appendix F (p. 236) and Procedure 6 (p. 99).

DAY 11

Record the results. Save mutants that appear to be insertions in *ompC* or *envZ*. In addition, save two insertions that appear to be near *ompC*. Prepare P1 lysates (see Procedure 10, p. 107) on these mutants and use them to transduce an *ompC-lac* fusion strain (MH225) to Tc^r (see Procedure 12, p. 111). Can *envZ*::Tn*10* be introduced into this strain? If so, this would provide evidence that the *envZ* gene does not specify an essential function. What is the phenotype of an *envZ* null mutation? Further mapping experiments will be conducted in Experiment 13 (p. 79).

EXPERIMENT 9
Targeted Mutagenesis of the Chromosome

Mutagenesis can be employed in bacterial genetics to increase the proportion of mutants in a given population. In cases where mutant isolation requires screening (Lac⁻ on an indicator plate) as opposed to selection (streptomycin resistance), mutagenesis is especially useful since it greatly decreases the number of colonies that must be examined to find the desired mutant. A number of different mutagens have been described (Miller 1983).

A major problem of these mutagenesis procedures relates to target size. If one mutagenizes the entire bacterial chromosome to obtain mutants in a particular gene, the probability of obtaining strains that have suffered multiple genetic lesions is high. A method designed to minimize this problem was suggested by Hong and Ames (1971). This method (local mutagenesis) is generally applicable, requiring only the knowledge of the chromosomal location of the target gene. Rather than mutagenize the entire chromosome, Hong and Ames mutagenized a lysate of the generalized transducing phage P22 (a *Salmonella typhimurium* generalized transducing phage) and then used this lysate in a transduction, selecting for incorporation of a donor gene near (i.e., cotransducible with) the region of interest. With this technique, only a small segment of the chromosome is mutagenized. Moreover, since each transductant will acquire a segment of DNA corresponding to the region of interest that has been heavily mutagenized, the probability of obtaining mutations in a gene located near the selective marker is high.

As described in Experiment 8 (p. 53), it is a relatively simple matter to insert a drug-resistance element near any gene of interest. Since this element can be used as the selective marker, the technique of local mutagenesis can be applied to any region of the chromosome. Indeed, drug-resistance elements allow one to mutagenize a target gene even if the chromosomal map position is not known.

There are now a number of different methods of local mutagenesis. One can either mutagenize generalized (Davis et al. 1980) or specialized (Experiments 11, 12, pp. 71, 75) transducing phages or mutagenize the donor bacteria prior to preparing a generalized transducing lysate (Hawrot and Kennedy 1976). The purpose of this experiment is to demonstrate the latter method.

The specific goal of this experiment is to use the technique of local mutagenesis to obtain a chain-terminating nonsense mutation (amber) in the *envZ* gene (see Appendix N, p. 283). For this purpose, we will mutagenize a donor strain that contains a Tn*10* insertion in *malP,Q* (*malP,Q*::Tn*10*). Since *malP,Q* is 50–60% cotransducible with *envZ*, by transducing and selecting for tetracycline resistance, we can focus on the region of the chromosome that contains *envZ*.

As discussed in Experiment 8, we do not know the phenotype that will be conferred by an *envZ* knockout mutation such as an *envZ* amber. Moreover, *envZ* may be an essential gene. In the previous experiment we avoid these problems by mutagenizing a strain diploid for *envZ*. In this experiment, since we are looking for *envZ* nonsense mutations, we will mutagenize a strain, SG263, carrying the nonsense suppressor *supF*. To score for *envZ* mutations, we utilize strain SG608 (Fig. 12) and again take advantage of the recessive nature of *envZ3*. The *envZ3* mutant phenotype, OmpC$^-$, can only be observed in haploids. The recipient strain (SG608) in the transduction is *envZ*$^+$ and carries the *envZ3* mutant allele on a lysogenic transducing phage. Mutations that destroy *envZ*$^+$ will unmask the *envZ3* mutant phenotype, OmpC$^-$. By using an *ompC-lac* fusion, the inability to express *ompC* can be scored as Lac$^-$. Since the recipient strain is diploid for *ompR*$^+$ and since we are mutagenizing only a selected region of the chromosome, the only expected OmpC$^-$ (Lac$^-$) mutations are those that lie in *envZ*. Suppressible nonsense mutations should be present at easily detectable frequencies. (If, in fact, an *envZ* knockout mutation is lethal, the only true knockout mutations that can be isolated will be amber mutations.) Once a series of characterized *envZ* nonsense mutations are isolated, the question of lethality can be directly answered.

Specific Comments

Students are often concerned because, following mutagenesis, cells are grown for several generations before the Pl lysates are prepared. They feel that this will increase the proportion of siblings present in the mutant population. (Remember, at least a generation of cell growth is required to segregate the mutation.) Although this is an important consideration when searching for spontaneous mutants, it is of little concern in this experiment. Mutagenesis is heavy, and the number of subsequent generations of cell growth is small. Accordingly, the number of siblings present is exceedingly small in comparison with the size of the total population.

This experiment yields Lac$^-$ mutants at a frequency $\geq 0.1\%$. Of these, 5–10% are suppressible nonsense mutations. Thus, the level of mutagenesis is very high; it is high enough so that mutants could be obtained even if this requires analysis of single clones by a biochemical method. Since mutagenesis is so efficient, the presence of closely linked multiple mutations must be seriously considered. This is especially true when nitrosoguanidine is used as the mutagen (Guerola et al. 1971). What experiments could be done to address this possibility?

In testing for phage K20 sensitivity (OmpF$^+$), particular care must be taken to ensure that the ratio of phage to cells is not too high (see Procedure 6, p. 99). If too many phage are used, resistant cells may appear sensitive.

DAY 1

Inoculate a 5-ml overnight culture of strain SG263.

DAY 2

Mutagenize SG263 with nitrosoguanidine (see Procedure 21, p. 129).

DAY 3

Dilute cells and prepare a P1*vir* lysate (see Procedure 10, Day 2, p. 107). Inoculate a 5-ml overnight culture of SG608.

DAY 4

Transduce strain SG608 to tetracycline resistance, selecting on lactose tetrazolium plates containing 8 µg/ml tetracycline (see Procedure 12, p. 111). Incubate plates at 37°C.

DAY 5

Score the number of Tc^r transductants. Determine the amount of phage lysate that gives the maximum number of transductants. If the optimum conditions yield more than 200 Tc^r transductants per plate, proceed as described below. If not, prepare a P1*vir* plate stock on the mutagenized culture (see Procedure 10, Notes). Prepare 15 separate transductions, using the optimal amount of lysate. Do not forget sterility controls.

DAY 6

Score results. If you are having trouble identifying Lac$^-$ transductants, try replica plating to lactose MacConkey agar containing 8 µg/ml tetracycline. Purify at least 30 Lac$^-$, Tc^r transductants on lactose tetrazolium agar containing 8 µg/ml tetracycline.

DAY 7

Repurify Lac$^-$ mutants on lactose MacConkey agar at 30°C.

DAY 8

Characterize Lac⁻ mutants by performing the following tests. Lac⁻ mutants derived from strain SG608 should be *envZ3*. They should exhibit these phenotypes:

1. pink on lactose MacConkey agar at 37°C;
2. white on lactose MacConkey agar at 30°C;
3. K20^s; test by cross-streaking (see Procedure 6, p. 99).

To score for amber mutations, mutant strains should be cross-streaked against φ80*p*SuIII on lactose MacConkey agar at 30°C. If the mutation is suppressible, the presence of the phage will confer a Lac⁺ phenotype at the intersection of the cross-streak.

DAY 9+

Score results. All mutants that appear to contain suppressible *envZ* nonsense mutations should be saved. P1 lysates should be prepared on these strains (see Procedure 10), and these lysates should be used to transduce an *ompC-lac* fusion strain (MH225) to tetracycline resistance (see Procedure 12). Can haploid strains containing a nonsense mutation be constructed? If so, this would provide evidence that the *envZ* gene does not specify an essential function. What is the phenotype of an *envZ* null mutation? Further mapping experiments will be conducted in Experiment 13 (p. 79).

EXPERIMENT 10
Isolating Chromosomal Deletion Mutations

Deletions are essential tools for genetic and biochemical analysis (see Experiments 1, 6, 13, pp. 7, 43, 79). Since they eliminate a region of DNA, the mutation is stable (i.e., nonreverting) and the phenotype conferred will behave in a predictable, recessive manner in diploid analyses. In addition, deletions remove genetic homology that could interfere with certain mapping experiments or biochemical methods involving hybridization.

In bacteria, chromosomal deletions occur spontaneously at low frequency. Historically, deletions have been isolated as mutations that simultaneously destroy the function of more than a single gene product. However, since many point mutations can have pleiotropic effects and since loss of gene function does not always confer a selectable or easily scorable phenotype, several procedures to enrich for deletion mutations have been developed.

One method that has been employed to isolate deletion mutations utilizes a transposable element such as Tn10 or bacteriophage Mu (see Appendix C, p. 226). These elements catalyze deletion formation to generate a series of deletions in a region of interest. For example, the drug-resistance element Tn10 carries a selectable genetic marker (tetracycline resistance, Tc^r) and can be inserted in any region of interest in the chromosome (see Experiment 8, p. 53). Using a media developed by Bochner et al. (1980) and improved by Maloy and Nunn (1981), one can isolate Tc^s mutants, which have lost the drug-resistance determinant. In this manner, deletions that extend into flanking chromosomal DNA can be obtained (Davis et al. 1980). However, transposable elements can catalyze inversions as well as deletions, and both events almost always leave a portion of the transposon (an IS element) at the novel joint in the chromosome. This can complicate further genetic and structural analysis.

The purpose of this experiment is to demonstrate a method for obtaining chromosomal deletions in a region of interest by exploiting the properties of gene fusions and the associated λ prophage (see Experiment 1). Almost 20 years ago it was observed that induction of a λ prophage integrated at the bacterial *att* site left a low frequency of surviving bacteria that had suffered deletions removing all or part of the λ DNA. Some of these deletions also extend into the adjacent chromosomal DNA (Calef and Neubauer 1969; Eisen et al. 1969; Pero 1971; Shimada et al. 1972). Since gene fusion technology enables one to insert a λ prophage at virtually any chromosomal site (see Experiment 1; Appendix L, p. 261), this method can be used to obtain deletions in any region of interest.

When λ exists as a prophage in *E. coli*, all its lytic functions are repressed by the cI gene product. When repressor is inactivated, λ transcription initiates at the powerful p_L and p_R promoters. When the N gene product is made from the p_L transcript, waves of transcription move through the prophage, overriding most transcription terminators. More prophage genes are expressed, including replication, excision, and lysis functions. Any cell experiencing prophage induction is doomed. This killing phenomenon has been studied in some detail and has led to many basic discoveries, including the technique we use in this experiment: the use of prophage induction to isolate chromosomal deletions.

Studies of prophage induction have been facilitated by use of a specific temperature-sensitive mutation in the cI gene called *cIts857*. With a *cIts857* prophage, one need only raise the temperature to induce the prophage in every cell in the culture. Cells surviving the heat treatment are easily found, and most contain deletion mutations that remove the lethal prophage functions.[1] Deletions are common because several prophage genes must be inactivated simultaneously. In general, prophage killing occurs because of N-mediated transcription, prophage replication, and expression of several so-called *kil* functions in the early operons of λ. If N-mediated transcription could be blocked, then neither *kil* nor replication functions would be expressed. However, transcription termination is naturally inefficient. Even if the prophage were defective in N function, some *kil* and replication functions are still expressed, leading to cell death. The simplest event that avoids killing by prophage induction is, in fact, a deletion removing the early operons of λ.

In this experiment we use the λ prophage that is adjacent to chromosomal gene fusions. Normally these phages carry a wild-type cI gene; however, we have devised a simple method to convert the cI^+ fusion prophage to *cIts857* (Fig. 13). The fusion phage can then be induced by simply raising the temperature. Since the site of integration of this phage into the bacterial chromosome was originally determined by a transposition event, no genetic homology or site-specific recombination functions of λ can mediate excision (see Appendix H, p. 241). Upon induction, events leading to cell death begin. Since the prophage cannot leave the chromosome, many rounds of both transcription and replication can extend into neighboring *E. coli* DNA. Mutants preexisting in the culture containing deletions that remove early prophage operons will survive heat induction. Some of these deletions will extend into the bacterial DNA. In addition, we suspect, although we have not proven, that some deletions are formed after heat induction, perhaps as a result of the extensive transcription and replication that occurs upon induction.

The specific goal of this experiment is to isolate a series of deletions that remove all or a portion of the *ompB* locus (see Appendix N, p. 283).

---

[1] If the prophage can leave the chromosome by recombination, most if not all of the temperature-resistant survivors will have simply lost the prophage.

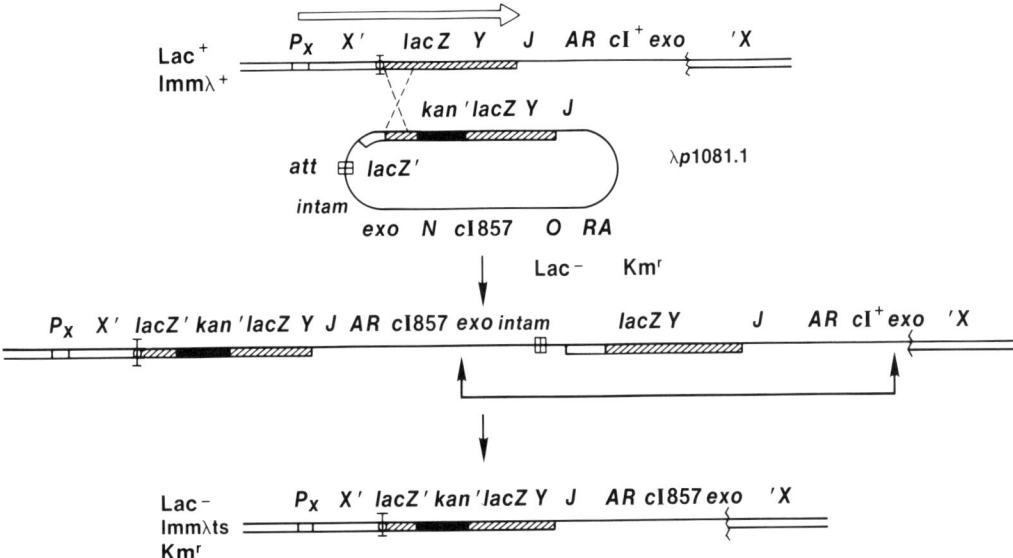

Figure 13
Use of λp1081.1 to convert a λcI⁺ lysogen to a λcIts857 lysogen. A lysogen of λp1081.1 is formed at the site of a *lac* fusion by selection for Km^r transductants. (This phage is *int*am, therefore the site-specific recombination system is inactive except in Su⁺ hosts.) This recombination is pictured in *lacZ* homology, although integration may also occur via λ homology. The phenotype of the double lysogen is Lac⁻ since λp1081.1 carries no promoter for the *lac* genes. However, since there is an intact *lacZ* gene, these lysogens are characteristically light blue on Xgal indicator medium. Segregation of one of the prophage genomes is shown occurring via λ homology. The resulting strain is a Lac⁻ λcIts857 lysogen. Since this lysogen does not contain an intact *lacZ* gene, it is white on Xgal indicator medium. A day-by-day protocol for these manipulations is given in Procedure 16 (p. 117).

This will be accomplished by selecting deletions in a strain with a *cIts857* prophage inserted at a gene fusion to *malP* (Fig. 14). This strain, SG158(pRT516.101), harbors a plasmid that carries *ompB* (Fig. 9, p. 34). The *ompB* region in this plasmid is *envZ⁺, ompR101*. This recessive *ompR* mutation confers an OmpF⁻, OmpC⁻ phenotype. The plasmid is required for the following reasons. First, a deletion of *envZ* from the chromosome may be lethal, and therefore *envZ⁺* must be provided. Second, the *ompR101* mutation on the plasmid is necessary to permit the scoring of deletions that enter the *ompB* locus. Obviously, if the plasmid carried both *envZ⁺* and *ompR⁺*, a deletion that removed the corresponding chromosomal genes would have no relevant phenotype. The *ompR101* mutation was chosen for this experiment, as opposed to *envZ3*, for example, because three-factor crosses have shown that *ompR* is closer to *malP* than *envZ* (see Experiments 8, 9, pp. 53, 59). Thus, all deletions that enter the *ompB* locus will confer a scorable phenotype, i.e., resistance to bacteriophage hy2.

To define the deletion endpoints within *ompB*, all those that enter the locus will be moved into an *ompC-lac* fusion strain by P1 transduction selecting for AroB⁺ (see Specific Comments below). Once the deletions are moved to this haploid background, we can determine by complementation tests if the deletion entered or removed *envZ*. This can be done by using a transducing phage, λpSG11, that carries *ompR*$^+$ but not *envZ*$^+$. If certain deletions cannot be moved into the haploid background, and if this transduction can only be done with deletions that are *envZ*$^+$, this would provide evidence that *envZ* is an essential gene.

Since deletions can extend in either or both directions from the λ prophage at *malP*, we will score for a variety of phenotypes as outlined in Day 4 of this experiment. This will permit us to define deletion endpoints. The order of the known genes in this region of the chromosome is *aroB, envZ, ompR, bioH, malQ, malP, malT, glpD, asd*. The phenotypes associated with the genes are:

> *aroB*: requires aromatic amino acids;
> *malQ,P*: inability to utilize maltose; sensitive to phage λ;
> *malT*: inability to utilize maltose; resistant to phage λ;
> *bioH*: requires biotin;
> *glpD*: inability to utilize glycerol aerobically;
> *asd*: requires diaminopimelic acid.

Specific Comments

This experiment utilizes strain SG158(pRT516.101). In SG158(pRT516.101) the λcI$^+$ fusion prophage has been converted to a λcIts857 prophage by the technique described in Figure 14. The techniques employed for this cross are not fundamentally different from those used in Experiment 1 (p. 7) to remove Mu sequences from Mu*d* fusions and replace them with λ.

The outside markers *aroB* and *asd* (Fig. 14) probably cannot be deleted from strain SG158(pRT516.101). Deletions that remove *asd* will require diaminopimelic acid, and this has not been provided in the selection media. To affect *aroB* may require that the deletion also remove a cluster of ribosomal RNA genes. This has never been observed and could indicate the presence of essential genes.

The phage used in this experiment for mapping, λpSG11, has the host range of φ80 because deletions that remove *malT* will be resistant to phages with the host range of λ.

Remember that the plasmid harbored by strain SG158(pRT516.101) carries more than just *envZ*$^+$ and *ompR101*. This plasmid carries ~8 kb of chromosomal DNA. Accordingly, the phenotype conferred by a particular deletion may be different in SG158(pRT516.101) than when the deletion is present in a haploid background.

It is possible to use the techniques described in Experiment 4 (p. 33) to prepare DNA from the haploid strains and physically determine the endpoints of deletions that affect *ompB*.

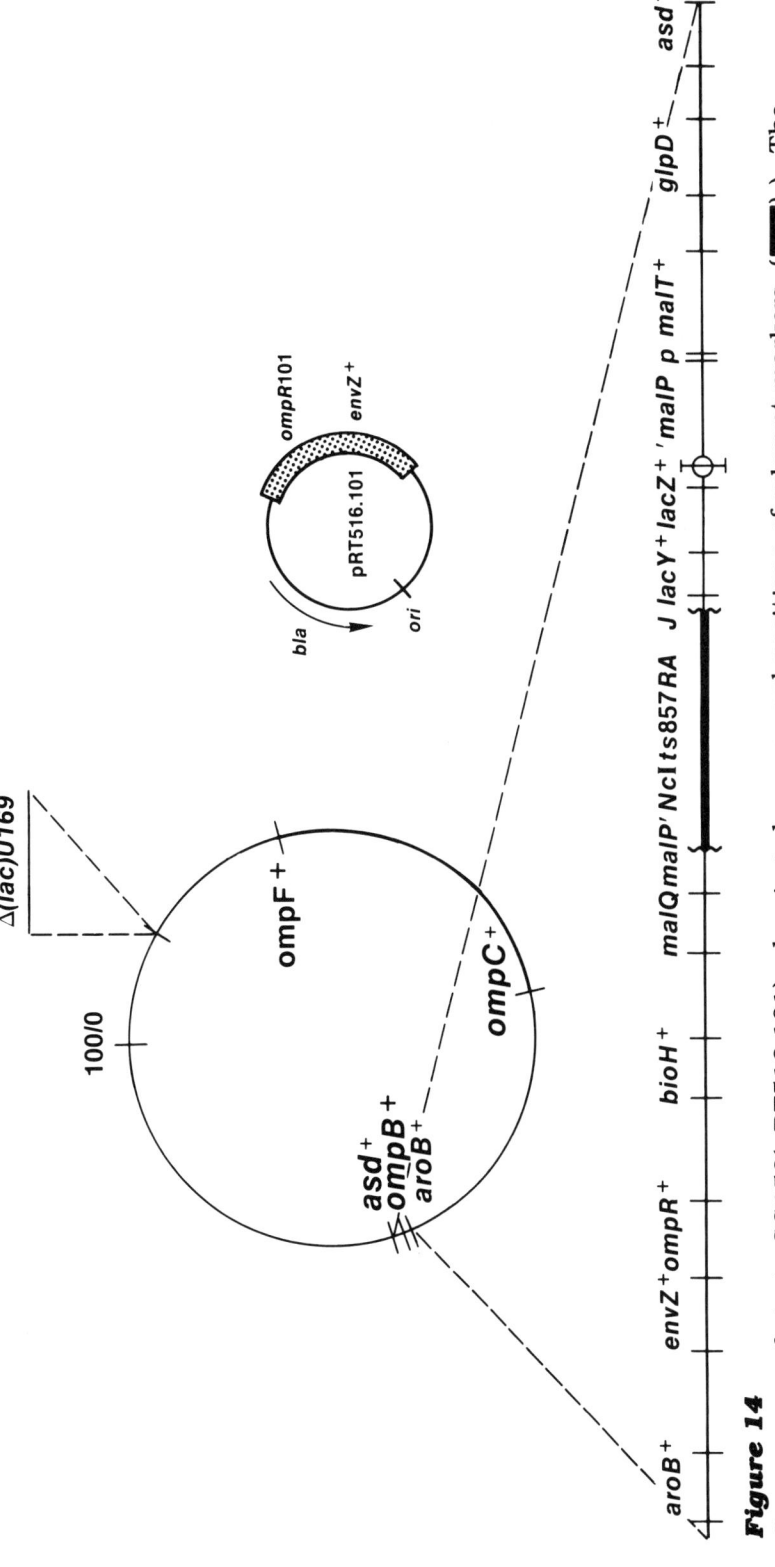

Figure 14
Genetic map of strain SG158(pRT516.101), showing chromosomal positions of relevant markers. (■) λ. The small circle at right represents plasmid pRT516.101, carrying a fragment from the chromosome (▩).

DAY 1

Streak strain SG158(pRT516.101) on L agar and incubate overnight at 30°C (see Procedure 18, Day 1, p. 121). Prepare 4 plates.

DAY 2

Continue Procedure 18 and pick single colonies to small test tubes containing 0.1 ml of L medium (1 colony/tube). Vortex the tubes and spread the contents on *prewarmed* (42°C) L plates. Prepare 20 plates. Also, try the patch method as outlined in the Notes to Procedure 18. One or 2 plates should be sufficient for this method. Incubate for 24–36 hr at 42°C.

DAY 3

Pick two or three colonies from each spread plate and one colony from each patch. Streak each isolate on L agar and incubate overnight at 42°C.

DAY 4

Repurify temperature-resistant mutants by streaking in L agar at 42°C.

DAY 5

Test each isolate by streaking or cross-streaking (see Procedure 6, p. 99) as follows:

1. L (master);
2. L containing kanamycin;
3. glycerol minimal medium;
4. phage hy2;
5. phage λ*vir*.

DAY 6

The tests from Day 5 will allow you to score for the loss of the genes *malT* and *ompR* and the kanamycin-resistance (Km^r) determinant on the λ prophage. Strains that do not grow on minimal glycerol will have lost the *bioH* gene, the *glpD* gene, or both. Since the purpose of this experiment is to isolate deletions that extend into, or beyond, the *ompB* locus, save only those strains that score as hy2^r (*ompR*). Furthermore, since genetic analysis may be complicated by deletions that leave any part of the λ prophage, hy2^r strains that remain kanamycin resistant can be discarded. Remember to record the phenotypes of each strain you retain for further analysis. These strains should be picked and streaked as follows:

1. L (master);
2. glucose minimal medium;
3. glucose minimal medium containing biotin;
4. glycerol minimal medium containing biotin.

Inoculate a 5-ml culture of SG626 and grow overnight at 37°C.

DAY 7

Score the phenotypes on the plates from Day 6 to define the deletion endpoints. Inoculate the deletion strains in 5 ml of L medium and make P1*vir* lysates as outlined in Procedure 10 (p. 107; they can be made the same day if you do not start too late in the day). Use these P1*vir* lysates to transduce strain SG626 to Aro$^+$ (growth on M63 glucose containing biotin) (see Procedure 12, p. 111).

DAY 8

Pick 8–16 Aro$^+$ colonies for each deletion strain and purify by streaking on the following media:

1. glucose minimal containing biotin;
2. lactose MacConkey;
3. maltose MacConkey.

DAY 9

From the minimal glucose plate, pick an isolated colony that is also Lac$^-$, Mal$^-$. These strains should now carry the deletions you isolated in strain SG158(pRT516.101). Inoculate a 5-ml culture of strain SG626 and its derivatives that carry the deletions and grow overnight at 37°C.

DAY 10

Centrifuge at 3000 rpm for 10 min and resuspend the pellet in 0.5 volume of 10 mM MgSO$_4$. In separate test tubes, mix 0.2 ml of each strain with 2.5 ml of molten (45°C) MacConkey top agar. Spread the mixture on a lactose MacConkey plate and let solidify. Using a sterile pasteur pipette, spot a drop of several dilutions of the λpSG11 phage on each plate. The phage λpRT2-80 should be used as a positive control. After the spots have dried, invert the plate and incubate at 37°C. These plates should be scored for Lac$^+$ plaques or papillae 5 hr from the start of incubation and periodically for 1–2 days. See Experiment 13 (p. 79) and Plate 2 for a more detailed description of this test.

EXPERIMENT 11
Isolating Deletion Mutations on a λ Transducing Phage

A useful feature of λ transducing phages and cloning vectors is the ease with which deletion mutations can be selected (Parkinson and Huskey 1971; Sternberg et al. 1979). As summarized in Experiment 10 (p. 63), deletions represent an important class of defined mutations. Transducing phages carrying these mutations can be employed to construct a genetic map (Experiment 13, p. 79). Since the physical analysis of λ DNA is straightforward, one can readily correlate the phenotypes conferred by different deletions with genome structure. Moreover, using the procedures described in Appendix J (p. 253), it is possible to recombine many of these mutations onto the chromosome to yield precisely defined bacterial mutants. The general purpose of this experiment is to demonstrate a simple technique for the isolation of deletion mutations in λ transducing phages.

Under proper conditions, treatment of λ with chelating agents, including EDTA or pyrophosphate, causes the phage heads to burst. This occurs because magnesium (Mg^{++}) ions are required to hold the capsomers of the head in place. DNA within the phage head exerts pressure that ruptures the structure when Mg^{++} ions are removed by the chelating agent. Under controlled conditions, phages resistant to chelating agents can be isolated. These resistant phages were found to contain deletions of phage DNA (Parkinson and Huskey 1971). It is possible to control the size of selected deletions by judicious choice of the concentration of the chelating agent, time of exposure, and temperature. λD69, like most λ cloning vectors, carries little nonessential λ DNA and is resistant to chelating agents. Insertion of a DNA fragment in λD69 increases the size of the phage genome, resulting in sensitivity to chelators. When the total size of the vector DNA plus insert DNA is approximately 50 kb, the phage is as sensitive to chelators as wild-type λ. With smaller phages, more-vigorous chelation treatment is required to isolate deletions. Because the nonessential λ DNA is removed in the λD69 vector, most deletions isolated by the chelator method will have at least one endpoint within the cloned fragment.

Deletions occur spontaneously during propagation of any λ stock. The frequency is generally about 1 plaque-forming deletion phage in 10^4 total phage (Parkinson and Huskey 1971). Occasionally, some transducing phages carry DNA sequences that are rather unstable and show unusually high deletion frequencies (Umene and Enquist 1981). Such instability may be due to expression of genes detrimental to λ or to unusual DNA structures, including direct repeats of DNA or active transposons.

Phenotypic variants of λ, called λ*, are resistant to chelating agents for reasons other than a lower DNA content. These λ* phages exist in any lysate at easily detectable frequencies. Consequently, several cycles of chelator selection are necessary to remove these phenotypically resistant variants. This is done by plaque purifying the phages several times in the presence of a chelating agent. Subsequently, the chelator-resistant phage can be analyzed for loss of genes on the transducing segment. In addition, the size of the deletion can be estimated by comparing the plating efficiency of a set of known deletion phages to that of the unknown deletion (see Procedure 19, Fig. 18, p. 125). A more precise estimate of deletion size can be made using CsCl density gradients (see Procedure 33, p. 160).

The specific purpose of this experiment is to isolate deletions in a λD69 hybrid phage carrying the *ompB* locus (see Appendix N, p. 283). This hybrid, called λpRT2 (Taylor et al. 1983), was isolated from an *E. coli* library constructed as described in Experiment 5 (p. 39), using the complementation methods described in Experiment 7 (p. 47). The EDTA-resistant phages will be isolated and screened for loss of *ompR* and *envZ* function by using specific gene fusion strains and lactose MacConkey plates. The sizes of the deletions will be estimated by testing plating efficiencies on EDTA plates and comparing these with known standards. DNA will be isolated from the EDTA-resistant phages and digested with several restriction enzymes to localize the deletions on the λpRT2 physical map. Phages that carry *ompB* deletions will be used in Experiment 13 to construct a genetic map.

Specific Comments

Examine the map of λD69 in Appendix I (p. 244) and note the location of the *exo* gene. The *gam* gene is to the right of *exo*, and there is an EcoRI site within *exo*. When deletions extend into this region, the usual result is phage that make smaller plaques. However, if the deletion removes both the *exo* and *gam* genes, the phage have a striking phenotype called Fec$^-$ (see Appendix F, p. 236). Unlike the undeleted parent, Fec$^-$ phage cannot plate with high efficiency on *recA* mutants. If you find such a deletion, it enters the cloned fragment from the right. If you have genetic data, this fact alone may aid in physically locating genes on the transducing segment. Phages with a Fec$^-$ phenotype can also be selected because they plate on *E. coli* P2 lysogens and Fec$^+$ phages will not. This phenotype is called Spi (see Appendix F). The Spi selection can be a useful mapping tool when coupled with EDTA selection (Umene and Enquist 1981).

The λD69 vector contains the *nin*5 deletion. This means that λD69 hybrids can carry deletions that remove the essential gene N and still grow (*nin* = N *in*dependent). You should not be surprised, therefore, to find an occasional clear plaque among the EDTA-resistant phages. This phage would be missing a number of λ genes including *red*, *gam*, N, and cI. What would its Fec and Spi phenotype be? What would be the plaque phenotype of a deletion that removed *exo*, *gam*, and cIII?

An alternative technique for isolation of deletions in λ relies on the use of an amber mutation in the capsid gene D (Sternberg et al. 1979). The technique is based on the finding that the normally essential product of gene D is dispensable for phage growth if the DNA content of the phage is less than 82% of wild type. A significant fraction of the rare phages that make plaques when a D amber mutant is plated on a nonsuppressing host contain deletions. A specific cloning vector, λDamsrIλ3, is designed to use the D amber selection method for isolation of deletions in the cloned fragment (see Procedure 20, p. 127; Appendix I).

DAY 1 *(late in the day)*

Start a 5-ml overnight culture of LE392 in TB containing 0.2% maltose at 30°C. Read Procedure 19 (p. 123).

DAY 2

Follow Procedure 19. Plate out single plaques of λpRT2 on LE392. Use TB plates and TB top agar. Spot dilutions of λpRT2 and the deletion phages on TB and TB-EDTA plates.

DAY 3

Examine the results from Day 2. How does the EDTA resistance of λpRT2 compare with that of the deletion phages? Prepare pickates from single plaques of λpRT2 grown on TB plates for EDTA selection (see Procedure 19).

DAY 4

Pick and purify EDTA-resistant plaques on TB-EDTA plates. After 4–6 hr, using toothpicks, stab EDTA-resistant plaques to LE392 lawns on TB plates (no EDTA).

Grow up 5-ml overnight cultures in TB of MH2101, SG624, and SE5000.

DAY 5

Harvest stabs. Begin genetic analysis by spotting pickates on lawns of MH2101 and SG624, using lactose MacConkey plates and MacConkey top agar. Test the pickates for their ability to plate on the *recA* strain SE5000.

DAY 6

Examine MacConkey spot-test plates and plaque purify any OmpB$^-$ phage.

DAY 7

Make 5-ml liquid lysates of selected phage deletions (see Procedure 1, p. 89).

DAY 8

Follow Procedure 27 (p. 142) to make λ DNA using 2 ml of the liquid lysates. Analyze the DNA with *Bam*HI, *Eco*RI, and *Cla*I. Include λ*p*RT2 DNA as a control. Save the remaining 3 ml of the lysates for use in Experiment 13 (p. 79).

EXPERIMENT 12

Targeted Mutagenesis of a λ Transducing Phage

In Experiment 9 (p. 59), a technique was described that allows targeted mutagenesis of a chromosomal region. Specialized transducing phages provide yet another means to target mutagenesis (see Experiments 1, 5, pp. 7, 39). These phages carry only a small fragment of chromosomal DNA. They can be readily isolated and mutagenized independent of any particular host-cell background. Accordingly, mutations that affect the target gene can be obtained following mutagenesis of the transducing phage by screening or selecting in an appropriate nonmutagenized recipient. Here again, only a small portion of the chromosome is exposed to the mutagenic agent.

λ fusion phages specifying hybrid proteins like those generated in Appendix L (Fig. 28, p. 262) provide a novel opportunity to isolate defined mutations in any target gene. Since translation of *lacZ* gene sequences requires translation of target gene sequences, we can obtain mutations in the target gene by screening for loss of β-galactosidase activity. Subsequent recombination of these mutations into the chromosome allows us to reconstruct the target gene carrying the defined mutation (see Appendix J, p. 253).

The general purpose of this experiment is to isolate a nonsense mutation in a target gene sequence fused to *lacZ*. This fusion is carried on a specialized transducing phage. This phage will be mutagenized and nonsense mutations that prevent synthesis of the hybrid protein will be identified by scoring the Lac phenotype of the phase in Su$^-$ and Su$^+$ hosts. Mutations that alter the mutant target gene and not *lacZ* will be identified by genetic mapping. Since *lacZ* can be fused to any gene (see Experiments 1, 2, pp. 7, 18; Appendix L, p. 261), this method can be employed even if the phenotype conferred by a mutation that abolishes the function of the target gene product (knockout mutation) is unknown or difficult to score.

We will mutagenize the fusion phage by using two different treatments. The first is a versatile in vivo technique based on the *E. coli* mutator gene, *mutD*. The second uses UV irradiation as a mutagen. Both methods are known to introduce a variety of mutations and therefore have a broad utility in bacterial and phage genetics.

The mutator allele *mutD5* increases mutation frequencies 10^3-fold to 10^4-fold when cells are grown in rich media (Fowler et al. 1974). The mutator effect is stimulated by thymidine and is reduced when cells are grown in minimal media. L broth and TB have sufficient thymidine for maximal mutagenesis. The *mutD5* allele promotes all possible base-pair substitutions; however, transitions occur more frequently than transversions. In addition, *mutD5* increases the frequency of base-pair additions (Cox 1976). Recently the *mutD* mutation has been shown to alter the editing capacity of DNA polymerase III (Echols et al. 1983).

The *mutD* system is particularly useful for mutagenizing phage because one need only grow the phage on a *mutD* strain to obtain a mutagenized stock (see Plate 3) (Enquist and Weisberg 1977). It is often sufficient just to make plaques on a *mutD* lawn. Each plaque usually contains enough mutants for most purposes. A plate stock of phage grown on *mutD5* will be a continuing source of independent mutants. Although lysates of chemically mutated phage are generally unstable and often toxic to the host cells, *mutD* lysates are typically as stable as normal lysates.

UV radiation is another broad-spectrum mutagen. The method used here is based on mutagenic repair of UV-damaged phage and is termed the Weigle effect (Weigle 1953). A phage stock is irradiated with sufficient UV light to lower the apparent titer by five orders of magnitude. This introduces lesions in the phage DNA, among these, thymine dimers. This stock is stable and is used to infect cells whose repair pathway (SOS pathway) has been induced by a brief exposure to UV.[1] The SOS system of *E. coli* recognizes the UV lesions in the infecting phage DNA and repairs the damage in an error-prone fashion, giving rise to phage mutants that appear when the infected cells lyse (Little and Mount 1982).

The specific goal of this experiment is to isolate amber mutations in the *ompR* gene. The mutations in *ompR* will be isolated initially in a phage carrying an *ompR-lacZ* fusion. Once isolated in this way, the mutations can be readily transferred to the *ompR* gene of the *E. coli* chromosome (see Appendix J).

We will use a specific *ompR-lacZ* protein fusion carried on the λ phage λTK10. The OmpR protein has a molecular weight of about 29,000 (29K). The OmpR-LacZ protein made by λTK10 contains about 25K of OmpR at the amino terminus and 110K of LacZ at the carboxyl terminus. This hybrid protein, which is active as β-galactosidase, is made under the control of the *ompR* promoter and uses the normal *ompR* translational start signals. When the fusion is expressed, the phage makes a blue plaque on Xgal plates. The essence of the experiment is to plate mutagenized λTK10 on lawns of MC4100 (an Su$^-$, *lac* deletion strain) on Xgal plates. Mutations that affect the synthesis or activity of the hybrid protein will give rise to phages that make white or pale blue plaques. Since approximately 75% of the hybrid protein is encoded by the *lacZ* gene, we expect that most of the phages making white plaques will carry mutations in this portion of the hybrid gene. However, 25% of the mutations should occur in *ompR* sequences. Phages carrying mutations in *ompR* can be recognized by their ability to transduce an Su$^-$, *lac* deletion strain to Lac$^+$. This transduction requires a specific recombinational event of the type described in Appendix J.

---

[1] This repair pathway can only be activated in *recA*$^+$ cells.

Specific Comments

A mutation frequency of 1% (i.e., clear plaques in a turbid phage or Lac⁻ plaques in a Lac⁺ phage) is indicative of good mutagenesis.

When working with transducing phages, it is possible (and sometimes desirable) to obtain recombination with the chromosomal allele. However, when you are isolating mutations in the transducing segment, you may lose the mutation by recombination during the process of purifying and making phage stocks. To eliminate this problem, one could do all the work by using an appropriate deletion mutant. If such a strain is not available, one can use *recA* strains as we do in this experiment (SE5000) to circumvent the problem of recombination.

It is important to use proper controls to verify mutagenesis.

DAY 1

Read Procedures 22 (p. 131) and 24 (p. 135). Late in the day, begin by inoculating 5-ml overnight cultures of LE30, 594, and SE5000.

DAY 2

Using a λTK10 lysate, follow mutagenesis procedures (*mutD*, Procedure 24; UV, Procedure 22). To ensure that you have good mutagenesis, titer *mutD* plate stocks and plate 0.1 ml from two UV mutagenesis tubes, using SE5000 lawns with Xgal indicator (see Procedure 8, p. 104).

Inoculate a 5-ml overnight culture of SE5000.

DAY 3

Assess mutagenesis (score clear plaques and white plaques as mutants). If there is a reasonable frequency of mutants, do a large-scale mutant hunt by plating out the mutagenized phage stocks on SE5000 lawns, using Xgal as indicator. Use 10 plates with 1000–2000 plaques per plate. (The titers of the UV mutagenesis tubes may not be high enough for more than 500 plaques per plate, plating out 0.5 ml.) The intention is to screen at least 10,000 mutagenized plaques.

Inoculate a 5-ml overnight culture of MBM7014 and SE5000.

DAY 4

Examine plates and, with toothpicks, stab all isolated white and pale blue plaques first to a TB top agar lawn of SE5000 (Su⁻) and then to a TB top agar lawn of MBM7014 (Su⁺) with Xgal in the top agar. Incubate at 37°C overnight.

DAY 5

Score presumptive nonsense mutants (white or pale blue on SE5000 and blue on MBM7014). Pick presumptive nonsense mutants from the SE5000 plate to 0.5 ml of L broth plus 0.05 ml of chloroform. Purify mutants by streaking or plating for single plaques on SE5000, using TB top agar containing Xgal. Incubate at 37°C overnight.

DAY 6

Early in the day, stab single plaques to TB top agar lawns of SE5000 to amplify each phage for testing. Incubate at 37°C for 6–8 hr or until lysis is visible around each stab. Use a 1-ml plastic pipette to pull the entire stab as an agar plug to 1 ml of λ-dil with 0.05 ml of chloroform (this phage suspension is called a pickate). Verify the presence of a nonsense mutation on SE5000 and MBM7014 lawns, as before, on Xgal plates.

It is critical to prove that the mutation is in $ompR$ and not $lacZ$. An unequivocal yet simple test is a transduction of MCM4100 to Lac$^+$. Do this by spotting an aliquot of each mutant plaque suspension on a lawn of MC4100, using lactose minimal plates and F top agar. The expected result of a simple $lacZ$ mutation would be no Lac$^+$ transduction since MC4100 is $\Delta(lac)$. However, λTK10 with an $ompR$ mutation should still transduce MC4100 to Lac$^+$. If these Lac$^+$ transductants simultaneously become OmpR$^-$ (hy2^r), any trivial explanation such as reversion or contamination of the phage is ruled out. Why? Draw out the recombination event. (If you are having trouble, consult Appendix K, p. 259.) What would you expect if the mutation were in $ompR$ but close to the fusion joint?

Inoculate a 5-ml overnight culture of SE5000 for plate stocks of interesting mutants.

DAY 7

Examine your plates and record results. Remember, our objective is to find a nonsense mutation in $ompR$. What other kind of $ompR$ mutations will give you a Lac$^-$ phenotype?

Do not discard the Lac$^+$ transductants of the potential $ompR$ mutations. These transductants are valuable in transferring the mutation from the fusion phages to the chromosome (see Appendix J, p. 253).

Make stocks of interesting mutants, using SE5000 (see Procedure 1 or 2, pp. 89, 91) and verify the phenotypes of the resulting stocks.

EXPERIMENT 13
Constructing a Genetic Map

Genetic analysis requires more than the isolation of mutants that exhibit an altered phenotype. Complementation tests must be performed to determine the number of genes required to confer a particular phenotype. In addition, the mutations must be mapped relative to each other and to other linked genetic loci to determine chromosomal location and gene order. Such an analysis may also provide important insights into the presence or absence of operons, the existence of important regulatory mechanisms, or distinct functional domains within a structural gene.

The general purpose of this experiment is to illustrate the principles and techniques involved in constructing a genetic map. In particular, we will construct a genetic map of the *ompB* locus (see Appendix N, p. 283). To do this, we will employ the chromosomal *ompB* insertion and point mutations isolated in Experiments 8 (p. 53) and 9 (p. 59) and the *ompB* deletion mutations isolated on a λ transducing phage in Experiment 11 (p. 71). Once again we will use strains carrying gene fusions to score results. Since OmpB⁺ is not an easily selectable phenotype, this mapping experiment would be extremely difficult to do using any other method.

The chromosomal *ompB* insertion and point mutations should be introduced into the *ompC-lac* operon fusion strain MH225. In this strain, the *ompB* mutations will confer a Lac⁻ phenotype, and this can be exploited to construct the genetic map. The various λ transducing phages that carry specific *ompB* deletions will be spotted on lawns of the different insertion and point mutations on lactose MacConkey agar. Lac⁺, i.e., OmpB⁺, will be seen as red lysogens growing out of the individual plaques (Plate 2).

The experiment is designed to show both complementation and recombination in a single test. If, for example, a mutation is present in the chromosome of the fusion strain that affects only *ompR*; and if the λ transducing phage carries a deletion affecting only *envZ*, each lysogen will be Lac⁺. In other words, when complementation is observed, all plaques will contain a uniform population of Lac⁺ lysogens (Plate 2). Such a result would indicate that the chromosomal point mutation and the deletion mutation on the λ transducing phage affect different genes. If the deletion mutation and the point mutation both affect the same gene, two possible results can be obtained. First, if the deletion and point mutations do not overlap (i.e., they affect different regions of the same structural gene), most lysogens will remain Lac⁻ (white). However, rare Lac⁺ recombinants will be observed in a portion of the plaques (Plate 2). Second, if the deletion removes the DNA corresponding to the point mutation, then a recombinational event yielding a Lac⁺ lysogen is impossible and, accordingly, all lysogens will remain Lac⁻. By crossing all of the point mutations with all of the deletions and noting which crosses yield Lac⁺ lysogens by complementation, which yield Lac⁺ lysogens by recombination, and which yield no Lac⁺ lysogens, we can deduce the position and extent of the various deletions relative to the point mutations. Careful compilation of this data should allow us to expand the simple *ompB* genetic map shown in Figure 15.

In genetic mapping experiments, the number of recombinants that exhibit a wild-type phenotype will increase in proportion to the distance between the two genetic lesions. Indeed, this frequency of recombinants can be used to position mutations relative to each other. This raises the following issues.

1. If the distance between two mutations is relatively large, the number of wild-type recombinants can be quite high. In our simple plaque test, this can cause confusion since it may appear that *all* lysogens are Lac⁺ when, in fact, a small proportion are not. From plaque appearance, this result could be mistakenly scored as complementation. This seemingly simple distinction has led many investigators astray. If you think that a particular phage complements a particular point mutation, streak from the center of a plaque onto lactose MacConkey agar spread with >10⁹ clear phage to select for lysogens. Are any of them Lac⁻? Remember, if it is true complementation, *all* lysogens will be Lac⁺. Ideally, complementation tests should be done in cells that are deficient in recombination (e.g., *recA*).

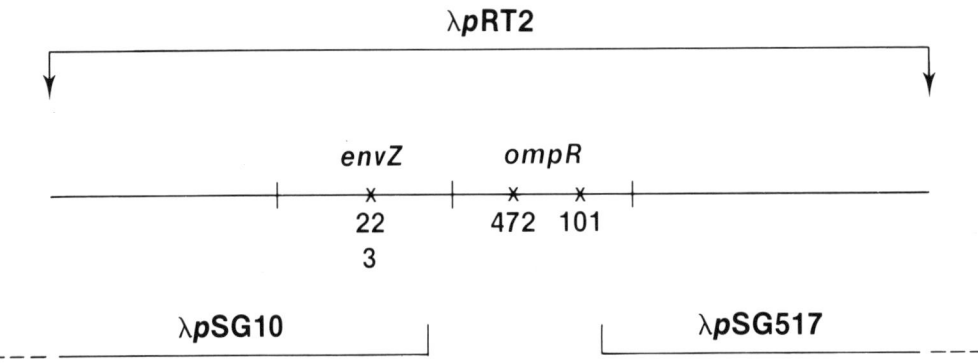

Figure 15
Genetic map of the *ompB* locus. (×) Point mutations in *ompR* and *envZ* (Hall and Silhavy 1981b). Arrow above the map shows the extent of the DNA incorporated into λpRT2 (Taylor et al. 1983). The two brackets in the lower part indicate the extent of material deleted from the indicated transducing phages. Phage λpSG10 was constructed by S. Garrett by cloning the *Bam*HI-*Bgl*II DNA fragment carrying *ompR* into the *Bam*HI site of λD69. Phage λpSG517 was isolated by S. Maloy and S. Fahnestock, using the techniques described in Experiment 11 (p. 71).

2. The frequency of recombination between two mutations is only a relative measure of genetic distance. A variety of factors that affect recombination in general can affect this frequency. Although it is tempting to order mutations based on the frequency of recombination, you should consider such data as tentative (Gross and Englesberg 1959).

3. In this experiment, we are crossing a deletion mutation with a point mutation, or, in other words, we are constructing a deletion map. If the deletion covers the point mutation, recombination to wild type is not possible. Accordingly, if wild-type recombinants appear, we can be certain that the deletion and point mutations do not lie in the same region of the DNA. Because of this certainty, a deletion map is considered to be quite reliable.

Specific Comments

To facilitate scoring, the following controls are provided (Fig. 15). All of the strains contain an *ompC-lacZ* fusion. In addition, all of these phages are *imm* 21 and will grow on the fusions strains that are λ lysogens.

Strain	Mutation
MH2101	*ompR101*
MH2472	*ompR472*
RT2003	*envZ3*
SG624	*envZ22*

Phage	Relevant bacterial gene carried
λpSG10	*ompR*$^+$
λpSG517	*envZ*$^+$
λpRT2	*ompR*$^+$ *envZ*$^+$

Mutants isolated in Experiments 8 (p. 53) and 9 (p. 59) and the mutant transducing phages isolated in Experiment 11 (p. 71) should be collected at this point. By using this collection, it should be possible to obtain a rather detailed genetic map of *ompB*.

Mapping *envZ* mutations can be difficult. Many EnvZ mutants are leaky and form pink colonies on lactose MacConkey agar. When a lawn is made from such a mutant, the color reaction is enhanced, and this reduces the contrast between a positive and negative reaction. Moreover, since EnvZ appears to be required in very low amounts, spot tests for complementation can be confusing. A more sensitive test would utilize lactose minimal agar instead of MacConkey; however, longer incubation times are required.

Recently, it has been demonstrated that *ompR* and *envZ* comprise an operon (Mizuno et al. 1982; Garrett et al. 1983). Since *envZ* is the distal gene, mutations isolated in Experiments 8 and 9 may be EnvZ$^-$ as a result of a polar mutation in *ompR*. Accordingly, do not be surprised if the complementation tests performed in the experiment reveal the presence of *ompR* mutations.

Experiment 13

DAY 1

Inoculate 5-ml cultures of MH2101, MH2472, RT2003, and SG624 as well as the *ompC-lac* fusion strains containing the mutations isolated in Experiments 8 and 9 (pp. 53, 59). Grow overnight at 37°C.

DAY 2

Centrifuge the overnight cultures at 3500 rpm for 10 min and resuspend the pellets in 2.5 ml of 10 mM Mg_2SO_4. Add 0.2 ml of each of the bacterial suspensions to small test tubes. To each tube, add 2.5 ml of MacConkey top agar, mix, and spread the contents onto lactose MacConkey plates. The phages λpSG10, λpRT2, and λpSG517 and the λompB deletion phages isolated in Experiment 11 (p. 71) should be diluted in λ-dil. Spot 10 μl of two dilutions containing approximately 50 and 500 plaque-forming units of each phage onto each agar plate. Make sure the plates are appropriately labeled according to the bacterial strain and position of the various *ompB* transducing phages. After the phage spots have dried, incubate the plates at 37°C for 8–16 hr.

DAY 3

Score results (see Plate 2). Continue to incubate the plates and score results for several days.

EXPERIMENT 14

Determining Gene Orientation by Using Gene Fusions to Isolate Specialized Transducing Phages In Vivo

For any gene, transcription proceeds in a 5'-to-3' direction from one of the DNA strands. Generally, regulatory signals (promoters, operators, ribosome-binding sites, etc.) are clustered at the 5' ends of bacterial genes. The order of mutations on the chromosome can be established (relative to known loci) by Hfr mating (Miller et al. 1968) or P1 transduction (Gross and Englesberg 1959) or by using the techniques described in Experiments 10 (p. 63) and 13 (p. 79). Although these methods can define the position of a gene as well as the order of various mutations at any locus, additional experiments are required to determine which mutations are "early" (promoter proximal), which mutations are "late" (promoter distal), and the direction of transcription relative to outside markers. The object of this experiment is to illustrate how *lac* gene fusions can be used to answer these questions. This experiment also illustrates how gene fusions can be used to obtain specialized transducing phages in vivo.

Once constructed, *lac* gene fusions can be used to determine the transcriptional orientation of any target gene. This type of analysis is possible because the methods used for in vivo gene fusion studies employ phages in which certain genes are always arranged in a fixed order relative to each other. More specifically, the transcriptional orientation of *lacZ* and *lacY* must be the same as the target gene, and the λ genes present on the adjacent prophage are in a fixed order relative to *lacZY*. Therefore, any experiment that can demonstrate the order of the prophage genes (including *lacZY*) relative to a known chromosomal locus will also demonstrate the transcriptional orientation of the target gene.

Wild-type λ integrates in a nonpermuted, unique orientation at a specific site on the *E. coli* chromosome (see Appendix F, p. 236). The genome of λ is 48.5 kb in length, and the position of the *cos* site is approximately in the middle of the prophage map. (The *cos* site is at the ends of the vegetative map, whereas the *att* site, which is at the ends of the prophage map, is in roughly the center of the mature phage map.) In the case of wild-type λ integrated at *att*λ, induction leads to *int*- and *xis*-promoted excision of the prophage. The frequency of illegitimate events leading to the release of phages that incorporate adjacent chromosomal genes is about 10^{-5}. These phages, some of which are defective, will transduce the *gal* or *bio* genes that are adjacent to *att*λ (Campbell 1971).

In Experiment 1 (p. 7), Lac⁺, plaque-forming transducing phages are isolated by UV induction of a fusion strain. Because the prophage in a fusion strain cannot be excised from the chromosome by any standard recombinational mechanism, this induction yields lysates whose titer of plaque-forming phages is only about 10^6/ml (see Appendix H, p. 241). These phages result from aberrant excision events that may include adjacent chromosomal genes resulting in the formation of transducing phages (Fig. 16a,b). In the same lysate there are also numerous defective transducing particles (Franklin 1971). Such defective, non-plaque-forming phages are the result of excision events that do not include all of the phage gene(s) necessary for plaque formation. The only requirements for defective phages are that the DNA be of proper size and carry the site necessary for packaging. In other words, these phages must be between 75% and 109% of λ size and must include the *cos* site that is located between the λ genes *A* and *R*. Of course, to propagate such defective transducing phages, a plaque-forming helper phage must be provided to complement the missing phage functions. In the original lysogen, presumably, these gene functions were expressed, allowing the proper packaging of the defective transducing phages.

In the case of a λ fusion phage that is "locked in" the chromosome, the aberrant excision events that occur upon phage induction yield at least two classes of defective transducing particles (Fig. 16c,d). One type of particle (Fig. 16c) will include genes to one side of the prophage, and the other transducing particle (Fig. 16d) will include markers from the opposite side. However, in all cases one arm of the λ prophage and all the markers present will be retained (e.g., either *lacZ* [Fig. 16d] or *imm*λ [Fig. 16c]). Therefore, by selecting for transduction of a linked chromosomal marker and by scoring for the presence of various prophage genes, the *orientation* of the target gene can be determined.

The specific goal of this experiment is to examine the lysates obtained in Experiment 1 for transduction of linked genes. These lysates are from

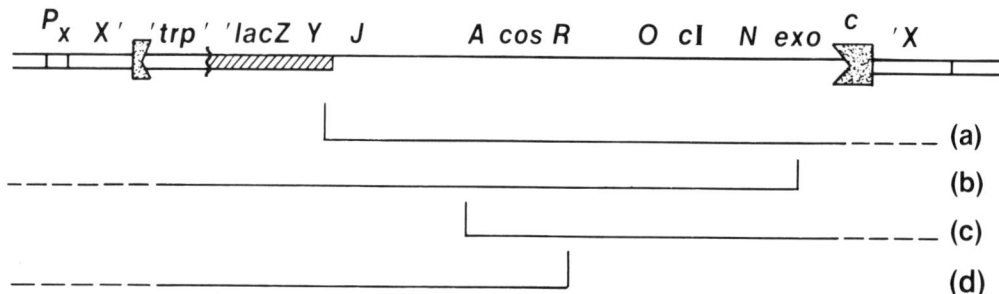

Figure 16
Genesis of a λ fusion transducing phage. The structure of an operon fusion and the adjacent λ prophage is shown (see Experiment 1, p. 7). The extent of DNA incorporated by four types of transducing phages is indicated by lines *a–d*. (– – –) Variable amounts of DNA that may be incorporated. The phage produced by events in lines *a* and *b* are plaque forming while those generated according to lines *c* and *d* will be defective. See text for details.

ara-lac fusion strains in which the prophage resides at the *ara* locus (min 1.2). We will examine these lysates for transduction of *leu* genes (min 1.7). By characterizing the specialized transducing phages present in these transductants, we can determine the orientation of the *ara* genes.

Specific Comments

Induction of a "locked-in" λ fusion phage yields a lysate, called a low-frequency transducing lysate (LFT), that contains many different transducing phages, each one present in small numbers. Using an LFT for transduction of *E. coli* markers, at least two types of transductants can be found. Generally, about one third of the transductants result from double reciprocal recombination events. The mutant allele is simply replaced by a wild-type gene in much the same manner as in the case of generalized transduction with bacteriophage P1. The other transductants are lysogens of the defective λ phage particles and therefore are diploid for the transduced marker. The diploid state, as well as the presence of λ genes, can be confirmed by various tests. For example, marker rescue and segregation by homologus recombination are two possibilities (see Appendix J, p. 253). Can you think of more? The original mutation (*leu*) is a Tn*10* insertion. What do you expect the phenotype of the transductants to be?

How much bacterial DNA can be incorporated in λ transducing phages? Since all infective particles must incorporate the *cos* site that is located at the midpoint of the prophage, then bacterial markers within 25 kb (or half the length of the phage genome) can be formed into transducing particles. Of course, the closer a marker is to the end of the prophage, the more frequently that marker will be incorporated in transducing phages. We can gauge the distance separating two markers by P1 cotransduction frequencies. There is a simple relationship between *E. coli* map distance in minutes (Bachmann and Low 1980; Bachmann 1983) and cotransduction of two genes (Wu 1966). Basically, the frequency of cotransduction is $(1-d/L)^3$ where L is the length of DNA in the transducing particle expressed in minutes of the chromosome, and d is the distance, again in minutes, between two markers. The capacity of the P1 particle is 100 kb, which is ~2.2 min (1 min = 45 kb). Assuming no effect of the specific markers on cotransduction frequencies, then two alleles 1 min apart will be 17% cotransduced. In practice, this frequency can range from a few percent to about 30%. Therefore, if the "reach" of a defective transducing phage is about 25 kb, then markers that show >42% linkage by P1 can, in theory, be incorporated into a defective transducing particle. In this experiment the recipient carries a *leu*::Tn*10* insertion mutation. The linkage of this Tn*10* insertion to *ara* is 80%.

Experiment 14

DAY 1

Inoculate 5-ml cultures of recipient strain MB100. Grow overnight at 37°.

DAY 2

Centrifuge the cultures and resuspend in 0.5 volume of $MgSO_4$. In small test tubes, mix 0.1 ml of cells with 0.1 ml of the λ lysates from Experiment 1, Day 11 (p. 17). Incubate for 10 min at room temperature for phage adsorption. Add 2.5 ml of molten (45°C) F top agar to each tube and pour on glucose minimal plates. Remember to include controls of the phage lysate alone and cells alone.

DAYS 3-4

Examine plates for Leu⁺ transductants. Purify candidates and check for:

1. *imm*λ, by cross-streaking against λcI (see Procedure 6, p. 99);
2. *lacZ*, by streaking on L plates spread with Xgal;
3. Tn*10*, by streaking on L plates containing tetracycline;
4. other phage genes according to Procedure 9 (p. 105).

DAY 5

Score results. In what direction are the *ara* genes transcribed?

PROCEDURES

PROCEDURE 1

Preparation of 2-ml High-titer λ Liquid Lysates

DAY 1 *(late in the day)*

1. Prepare a culture tube with 5 ml of TB.

2. Inoculate with a single colony of an appropriate host strain and shake or rotate at 30°C overnight.

3. Prepare single plaques of the desired phage by either streaking or plating out dilutions of a phage lysate or pickate in soft agar overlays (see Appendix G, p. 239).

DAY 2 *(early in the day)*

1. Centrifuge the overnight culture at 1500g for 10 min. Resuspend the cell pellet in 2.5 ml of 10 mM $MgSO_4$.

2. To five culture tubes, add 0.05 ml of the cells in $MgSO_4$.

3. Remove individual, well-isolated plaques as agar plugs by using a sterile pasteur pipette. Add the plaques to the test tubes prepared in the previous step as follows:

Test tube	No. of phage plaques
1	1
2	2
3	3
4	4
5	0

4. Vortex the tubes to mix phage and cells. Make sure the plaque is immersed in the cell droplet and incubate at room temperature for 5 min.

5. Add 2 ml of L medium containing 10 mM $MgSO_4$ and shake or rotate vigorously at 37°C for 4–6 hr or until lysis occurs.

6. As lysis occurs, remove the tube and add 0.1 ml of chloroform. Vortex to mix well and put on ice.

7. By 6 hr, all tubes except tube 5 should be lysed. If not, add chloroform and proceed. (Do not despair; high-titer lysates are occasionally obtained from tubes with no obvious lysis.)

8. Transfer the lysates to sterile centrifuge tubes and centrifuge at 4500g for 10 min. Pipette the supernatant to a sterile, screw-capped tube. Be sure to label and date your lysates.

9. Spot-titer the lysates (see Appendix G). Store the tubes in the dark at 4°C. Save the tube with the *highest* titer and the *lowest* number of input plaques. Ideally, tube 1 is the best. Titers should be near 10^{10}/ml or better. Discard lysates of 10^8/ml or less. A titer of 10^9/ml is barely adequate.

Notes

Phage λ is very sensitive to trace amounts of detergent, so make sure glassware is thoroughly rinsed free of soap. In addition, λ lysates are sensitive to visible light and should be stored in the dark.

PROCEDURE 2
Preparation of Phage Plate Stocks

DAY 1 *(late in the day)*

1. Prepare a culture tube with 5 ml of TB.

2. Inoculate with a single colony of an appropriate host strain and shake or rotate at 30°C overnight.

3. Prepare single plaques of the desired phage by either streaking or plating out dilutions of a phage lysate or pickate in soft agar overlays (see Appendix G, p. 239).

DAY 2 *(early in the day)*

1. Centrifuge the overnight culture at 1500g for 10 min. Resuspend the cell pellet in 2.5 ml of 10 mM MgSO$_4$.

2. Mix 10^5–10^6 phage with 0.1 ml of cells in a culture tube. If you have fresh, single plaques available, use 1–5 plaques per 0.1 ml of cells. In this case, pick plaques as agar plugs directly to 0.1 ml of cells.

3. Incubate at room temperature for 5 min. Be sure that the cells and phage are mixed well.

4. Add 2.5 ml of L medium followed by 2.5 ml of molten (45°C) top agar (see Notes below) and pour over a freshly prepared but solidified L plate.

5. Incubate *right side up* at 37°C in a wet box. (A wet box is a closed container with water-saturated paper towels in the bottom. This ensures wet plates, which, in turn, promote large plaques. A plastic box 8 in.×12 in. and at least 6 in. deep is sufficient. Wet boxes are not essential for good stocks; they just ensure regular success.)

6. The plates are very sloppy, so use care when examining them. Plaques should appear in 4 hr and spread to confluent lysis in 5–7 hr. When just confluent, scrape the soft agar overlay into a 15-ml Corex centrifuge tube.

7. Wash the plate with 2 ml of λ-dil and combine with soft agar in the centrifuge tube. Add 0.1 ml of chloroform and vortex.

8. Centrifuge at 4500g for 10 min to remove cell debris. Pipette the supernatant to a sterile, screw-capped tube. Label and date each lysate.

9. Titer phages as described in Appendix G. Store lysates, protected from light, at 4°C.

An alternative harvesting method is as follows:

1. In step 6 above, after confluent lysis, cool the plates by placing them right side up at 4°C.

2. When cold, overlay gently with 5 ml of ice-cold λ-dil. Store right side up at 4°C overnight.

3. On the next day, remove the liquid overlay to a centrifuge tube and add 0.1 ml of chloroform. Vortex briefly.

4. Remove the debris from the overlay fluid by centrifugation at 4500g for 10 min. Label and date the lysates. Titer the phage as described in Appendix G.

Notes

If you intend to make DNA from this plate stock, use 0.7% agarose in water rather than TB top agar. Impurities in agar tend to give DNA stocks that cannot be digested with restriction enzymes.

A supplemented mixture reported to give higher-titer plate stocks can be added to the bottom L agar (Day 2, step 4). Just prior to pouring the bottom agar, adjust to final concentrations of 0.3% glucose, 75 μM $CaCl_2$, 4 μM $FeCl_3$, and 2 mM $MgSO_4$ (Y. Nozu, pers. comm.).

PROCEDURE 3
Preparation of 1-liter λ Lysates

DAY 1 *(late in the day)*

1. Add 40 ml of TB containing 10 mM MgSO$_4$ to a sterile 125-ml flask.

2. Inoculate with a single colony of an appropriate host strain and grow with shaking overnight at 30°C.

3. Titer the phage lysate to be used (see Appendix G, p. 239).

DAY 2 *(early in the day)*

1. Mix 20 ml of fresh overnight cells (no centrifugation necessary) with about 5×10^8 phage in a 4-liter flask. This amount of phage is somewhat arbitrary. As a general rule, somewhere between 10^8 and 10^9 phage per 20 ml of overnight culture should suffice.

2. Incubate for 5 min at room temperature for phage adsorption.

3. Add 1 liter of L medium containing 10 mM MgSO$_4$. If you use media prewarmed to 37°C, the chances for a good lysate improve.

4. Shake vigorously (the more aeration, the better) at 37°C. You really cannot aerate too much. However, watch out for water splashing into the flask and flask liquid splashing out.

5. Lysis should be evident in 5–8 hr, depending on the phage. In general, incubation longer than 10 hr gives poor results. Lysis can be more clearly seen by looking at a 1-ml sample in a small, glass test tube.

6. After lysis occurs, add NaCl to 0.5 M (2.9 g/100 ml) and add about 1 ml of chloroform.

7. Continue shaking for 5 min.

8. Centrifuge the lysate at 6000g for 10 min to remove cell debris. Decant and save the supernatant. Label, date, and titer the lysate; store the lysate at 4°C.

9. Concentrate the phage for DNA purification (see Procedure 26, p. 140).

PROCEDURE 4
Rapid Method for Purifying Phage from Plate Stocks or Small Liquid Lysates

I. CENTRIFUGATION OF PHAGE

1. Phage can be purified quickly by pelleting through a glycerol step gradient in an ultracentrifuge. Use a Beckman 50Ti rotor (tube capacity=13.5 ml), SW41 rotor (tube capacity=13.2 ml), or equivalent. To the appropriate centrifuge tube, add 40% glycerol, 5% glycerol, and lysate in order according to Figure 17. Fill with λ-dil, if necessary. Centrifuge at 35,000 rpm for 60 min at 4°C.

2. Pour off and discard the supernatant.

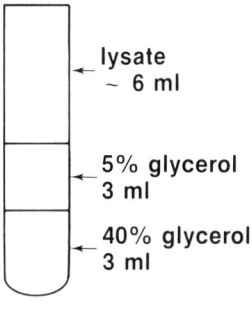

Figure 17

II. ISOLATION OF DNA

METHOD A: *Rapid Extraction*

1. Resuspend the phage pellet in 1 ml of λ-dil.

2. Add DNase I and RNase A to a final concentration of 1 µg/ml and 10 µg/ml, respectively (see Appendix A, p. 217).

3. Incubate for 30 min at 37°C. This sample is ready for lysis using a STEP buffer (see Procedure 26, steps 13–19, p. 141) or by phenol extraction (see Procedure 40, p. 177).

METHOD B: *CsCl Float-up*

1. Resuspend the phage pellet in 0.5 ml of 5.65 M CsCl ($\rho=1.7$, 25°C) in TM (10 mM $MgSO_4$, 50 mM Tris·HCl, pH 7.4). Add to the bottom of a Beckman SW50.1 centrifuge tube (tube capacity=5.0 ml) or its equivalent.

2. Overlay with 3 ml of 4.85 M CsCl ($\rho=1.6$, 25°C) in TM.

3. Overlay with 1 ml of 3.26 M CsCl ($\rho=1.4$, 25°C) in TM.

4. Centrifuge at 30,000 rpm for 1 hr at 20°C.

5. Locate the phage band with a microscope light. From the top, using a pasteur pipette with a bulb, *first* remove the liquid down to the band, *then* remove the band in a volume of not more than 0.5–1 ml. Use a microscope light to monitor the procedure.

6. Remove CsCl by dialysis and proceed with DNA extraction (see Procedures 26, 40, pp. 140, 177).

Note

Part I of this procedure can be scaled up to accommodate larger volumes of lysates.

PROCEDURE 5
Red-plaque Test for λ Int and Xis Functions

DAY 1

1. Prepare a culture tube with 5 ml of TB.

2. Inoculate with a single colony of red-plaque tester strain, LE292, and shake or rotate at 30°C overnight.

3. Prepare fresh galactose TTC plates.

DAY 2

1. Prepare single plaques of the desired phage by either streaking or plating out dilutions of a phage lysate or pickate in TB top agar overlays containing LE292 cells on galactose TTC agar.

2. Incubate the plates at 30°C or 37°C. Use the temperature that gives turbid plaques. Incubate until plaques and a color reaction develop. This takes 30–48 hr at 30°C or 12–24 hr at 37°C.

Notes

This is a plaque-color screen for the presence of λ integration-excision functions (Enquist and Weisberg 1976). λint^+xis^+ phage make red plaques on special *E. coli* strains using galactose indicator plates. Phage with a defect in either *int* or *xis* form white plaques. This test gives direct assessment of the state of the Int and Xis functions. One practical use is that λ vectors with cloning sites in *int* (e.g., the *Bam*HI site in λD69; Appendix I, p. 244) can be plated on red-plaque tester strains to determine if there are any DNA inserts cloned at the *Bam*HI site.

The test works as follows: Red-plaque tester strains carry a cryptic λ prophage integrated in the *galT* gene of the *gal* operon. The cryptic phage contains a large deletion from within the *int* gene to the *F*II gene. Consequently, there is virtually no λ DNA remaining except the prophage attachment sites. Such cells are Gal⁻ due to the prophage insertion. Infection of these strains with wild-type λ provides *int* and *xis* gene products that excise the cryptic prophage, restoring the cells to Gal⁺. Mutant phages (either *int* or *xis* or both) are recognized by their inability to promote this specific excision. The red color comes from the dye tetrazolium (2,3,5-triphenyl-2H tetrazolium chloride) in TB agar plates containing 1% galactose. The lawn remains white, and the center of turbid plaques are red for Gal⁺ (excision) and white for Gal⁻ (no excision); see Plate 3 at the front of the book. Phage that form clear plaques do not work well in this test.

PROCEDURE 6
Selection of λ Lysogens

DAY 1

1. Prepare culture tubes with 5 ml of L medium.

2. Inoculate with single colonies of each strain to be lysogenized and shake or rotate at 30°C overnight.

DAY 2 *(early in the day)*

1. Centrifuge the overnight culture at $1500g$ for 10 min. Resuspend the cell pellet in 2.5 ml of 10 mM $MgSO_4$.

2. In a small test tube, mix 0.1 ml of cells and 2.5 ml of molten (45°C) TB top agar. Quickly pour the mixture on TB plates; the soft agar overlay will solidify in a few minutes.

3. Spot the soft agar surface for each strain with three 50-μl aliquots of the λ lysate: undiluted, diluted 10^{-2}, and diluted 10^{-4}. Allow the spots to dry on the agar by leaving the plates on the bench with covers partially off.

4. Do not invert the plates. Incubate overnight at 30°C or 37°C (see Notes below).

DAY 2 *(late in the day)* **or DAY 3**

1. Spread L plates with 10^9 selector phage (see Notes below) and let them dry for a few minutes. Usually no more than 0.2 ml can be spread and dried easily.

2. Examine the plates prepared early on Day 2. Spots of lysis should be evident.

3. Using a sterile, flat toothpick, pick some cells from the turbid centers of lysis. Streak for single colonies on the plates spread with selector phage. Try to do eight sectors per plate. It is best to pick from spots of confluent lysis produced by the *least* number of phage.

4. Incubate overnight at 30°C or 37°C (see Notes below).

DAY 3 or DAY 4

1. Select large, round, healthy-looking colonies as potential lysogens and restreak on L plates to purify.

DAY 4 or DAY 5

1. Cross-streak the lysogens on TB plates against λcI and λvir (see Fig. 18). First apply a thin line of each phage lysate (~10^8/ml) with a 0.1-ml glass pipette. Allow to dry. Next streak a single colony or cell suspension across the dried phage lysate. Positive and negative cell controls must be included on every plate to ensure that the proper ratio of phage to cells has been used. If you cross-streak against both λcI and λvir in one streak, always streak cultures across the λcI first and then across the λvir in a continuous motion.

2. Incubate overnight at 30°C or 37°C (see Notes below).

DAY 5 or DAY 6

1. Select colonies that are λvir^s and immune to λcI. These are lysogens.

Notes

When spotting phage lysates for lysogen formation, use 30°C if the λ repressor is temperature-sensitive; otherwise use 37°C.

The selector phage for *imm*λ lysogens is λb2cI. To obtain lysogens with a different immunity, a different selector phage must be employed. For example, the selector phage for *imm*21 lysogens is λint6red3imm21c.

To select rare lysogens, it is often necessary to use a mixture of selector phage with different host ranges. A common mixture is λb2cI and λh80Δ(*int-red*)9cI. This approach greatly reduces the background of phage-resistant nonlysogens. However, it is important to check putative lysogens for the presence of selector phage markers.

Figure 18 shows an example of a phage-sensitivity test by cross-streaking. Three strains are shown: a λ^s strain (top), a λ^r strain (bottom), and λ lysogen (middle). The phages are λb2cI and λvir.

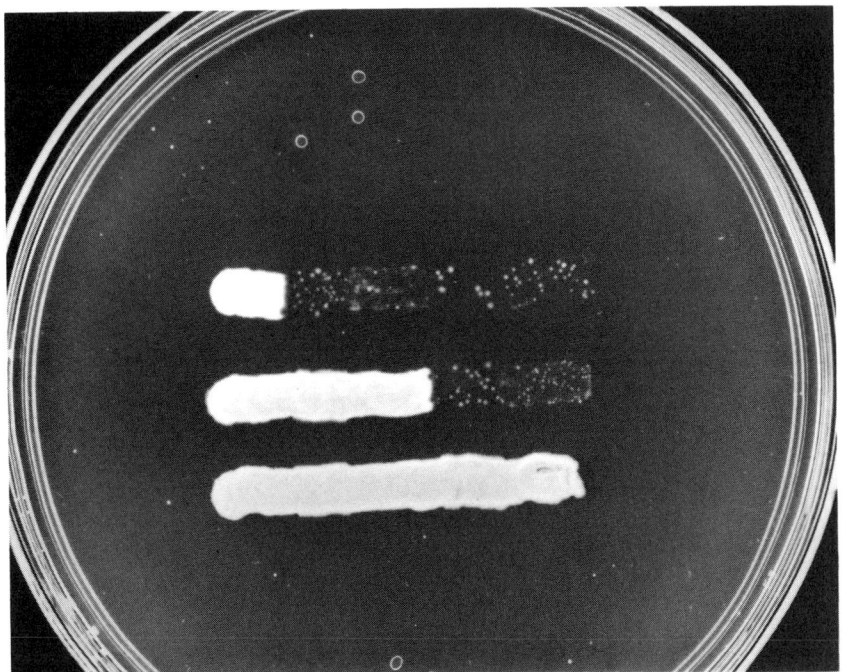

Figure 18
Cross-streak test for λ lysogens. This test is done as follows: On an L plate, ~50 μl of a λcI lysate (10^8/ml) is spread in a vertical line in the left-hand portion of the plate. This can be done by using a pasteur pipette. Parallel to this and to the right, ~50 μl of λvir lysate (10^8/ml) is spread in a similar line. The plate is allowed to dry, and consequently these lines are not visible in this photograph. Strains to be checked for lysogeny are spread perpendicular to and across the lines of both phages in a single swath from left to right. In this photograph the strain in the top streak is a nonlysogen, the strain in the middle is a lysogen, and the strain at the bottom is phage resistant.

PROCEDURE 7
Induction of λ Lysogens by Ultraviolet Light

1. Prepare a culture tube with 5 ml of L medium.

2. Inoculate with a single colony of a lysogen.

3. Grow at 37°C with aeration to a cell density of 2×10^8 to 4×10^8 cells/ml (3–4 hr). If the lysogen is temperature-sensitive, then growth must be at 30°C and it will take 6–8 hr to reach 2×10^8 to 4×10^8 cells/ml.

4. Centrifuge each 5-ml culture at $1500g$ for 10 min. Resuspend the cell pellet in 2.5 ml of 10 mM $MgSO_4$.

5. Pipette the cell suspension into an empty, sterile petri dish. Keep the dish covered.

6. Add 4.5 ml of L medium to an amber (or foil-covered) culture tube.

7. Expose the cell suspension in the petri dish to a dose of 350 erg/mm² UV irradiation. Remember to *remove the cover* from the petri dish.

8. Immediately transfer 0.5 ml of cells to the culture tube prepared in step 6.

9. Shake or rotate at 37°C for 2–5 hr until the cells lyse.

10. Add 0.1 ml of chloroform and vortex well to mix.

11. Transfer to centrifuge tubes and remove debris by centrifugation at $4500g$ for 10 min.

12. Transfer the supernatant to sterile, screw-capped tubes for storage. Be sure to label and date the lysates.

13. Titer the lysates as described in Appendix G (p. 239). Store the lysates at 4°C.

Notes

There is a simple and convenient method for UV induction of many λ lysogens on a single plate. This involves preparing a lawn of λ^s cells in TB top agar on L plates. Once the agar has solidified, colonies of lysogens are picked and stabbed into the lawn. The entire plate is then exposed to an inducing dose of UV irradiation. Following overnight incubation, zones of lysis should be evident around each stab. These areas can be harvested by using a 1-ml plastic pipette to remove an agar plug to a convenient volume of λ-dil. Single plaques from these pickates can be isolated and purified (see Appendix G).

PROCEDURE 8
Scoring LacZ+ Phage Plaques with Xgal

1. Make a solution of 20 mg/ml Xgal in N-N-dimethyl formamide (see Appendix A, p. 217). Be careful! This solvent dissolves plastic. Store, protected from light, at 4°C.

2. Titer the phage as described in Appendix G (p. 239), but be sure to use a *lacZ* (preferably a deletion) strain for the lawn. Add 0.2 ml of Xgal solution to the overlay agar suspension just before pouring on TB plates.

3. Incubate overnight at the desired temperature and score blue plaques as LacZ+ and white plaques as LacZ−.

Notes

Xgal solutions *cannot* be added directly to cells. Add molten top agar to the phage and cells and mix before adding Xgal.

On those bad days when you forget to add Xgal to the plates, it is often possible to rescue the experiment by letting the plaques develop and then pouring 2.5 ml of soft agar containing 0.2 ml of Xgal *over* the plaques. After a few minutes, the free β-galactosidase in the plaques of LacZ+ phages catalyzes the hydrolysis of the Xgal, turning the plaques blue. This is *not* the method of choice, however.

PROCEDURE 9
Detection of Phage Genes in Prophage Deletions

DAY 1 *(late in the day)*

1. Prepare a culture tube with 5 ml of TB.

2. Inoculate with single colonies of each temperature-resistant strain harboring potential prophage deletions (see Procedure 18, p. 121). Include the parental lysogen as a positive control and the nonlysogen as a negative control. Another positive control would be a strain with an amber suppressor, e.g., LE392 or MBM7014.

3. Shake or rotate at 30°C overnight.

DAY 2

1. Prepare a lawn of each culture to be tested. To a small test tube, add 0.1 ml of cells and 2.5 ml of molten (45°C) TB top agar. Quickly pour the cell suspension on a TB plate.

2. After the top agar overlay has solidified, spot 50-μl droplets of each of the four phage lysates in the tester set (see Notes below). Each plate should have four spots the size of a dime, and each spot should be uniquely labeled on the plastic underside of the dish. Each spot contains 10^6-10^7 particles of a particular λ amber mutant or λ wild-type control.

3. Allow the spots to dry on the agar by leaving the plate on the lab bench with the cover partially off. When dry, incubate at 37°C overnight (do not invert the plate). If the parental prophage is *c*Its857, incubate deletion candidates at 42°C.

106   Procedure 9

DAY 3

1. Examine the spots for lysis. The positive control is wild-type λ: All strains should show a distinct spot of lysis. The three amber mutant spots should show no, or only a few, plaques on the nonlysogen and confluent lysis on the parental lysogen. Confluent lysis on the putative deletion strains indicates the presence of that particular λ gene in the lysogen. No plaques or only a few indicates that a deletion extends into this particular λ gene in the prophage. One cautionary note: The tester-set phages are all *imm*λ. We have assumed that the prophage in the strains with potential deletions is heteroimmune or that the cI gene is deleted or inactivated (by growth at 42°C of *c*Its857). What would you expect if the original prophage were homoimmune?

Notes

The tester set for detection of prophage genes should contain lysates (10^8/ml) of specific λ phages carrying amber mutations in three widely separated genes as well as a lysate of λ*c*Its857. The specific amber alleles are:

1. λ *N*am7 *N*am53 *c*Its857 (a double amber mutation in positive control gene *N*).

2. λ *c*Its857 *S*am7 (a single amber mutation in lysis gene *S*; *S*am7 can only be suppressed by *supF*);

3. λ *W*am403 *c*Its857 (a single amber mutation in a gene that specifies a protein involved in phage morphogenesis).

   This rather simple experiment can yield results showing the more subtle aspects of λ genetics. A more detailed discussion of genetic mapping is presented in Experiment 13 (p. 79).

PROCEDURE 10
Preparation of P1*vir* Lysates

DAY 1

1. Prepare a culture tube with 5 ml of L medium.

2. Inoculate a single colony of the donor strain and shake or rotate at 37°C overnight.

DAY 2

1. Inoculate 0.05 ml of the overnight culture in 5 ml of L medium containing 0.2% glucose and 5 mM $CaCl_2$.

2. Incubate for 30 min at 37°C with aeration.

3. Add 0.1 ml of a P1*vir* lysate ($\sim 5\times10^8$ phage/ml).

4. Shake or rotate at 37°C for 2−3 hr until the cells lyse.

5. Add 0.1 ml of chloroform and vortex.

6. Centrifuge at 4500*g* for 10 min to pellet the debris.

7. Carefully transfer the supernatant to a sterile, screw-capped tube. Add 0.1 ml of chloroform and vortex to mix. Store the lysates at 4°C.

8. Spot-titer the lysate (see Appendix G, p. 239). Use L top agar containing 5 mM $CaCl_2$ and 10 mM $MgSO_4$ and L plates.

Notes

If cells do not lyse after 3−4 hr, proceed with step 5. Cultures that do not clear often yield usable phage lysates.

The most common mistake made when making P1 lysates is to start with a culture that is too dense. If this is done, the culture will not lyse.

P1 lysates are sensitive to chloroform if Ca^{++} is omitted. In addition, Ca^{++} is required for absorption of P1.

P1 can also be grown in plate stocks. This is done as described for λ (see Procedure 2, p. 91). Use approximately 10^7-10^8 phage and remember to use top agar containing 5 mM $CaCl_2$ and 10 mM $MgSO_4$ and L plates.

PROCEDURE 11

Preparation of a P1Tn9clr100 Lysate

DAY 1

1. Prepare a culture tube with 5 ml of L medium.

2. Inoculate a single colony of the P1Tn9clr100 lysogen and shake or rotate at 30°C overnight.

DAY 2

1. Inoculate 0.05 ml of the overnight culture in 10 ml of L medium.

2. Incubate at 30°C with aeration until the cells reach early-log phase. This corresponds to an OD_{600} of ~0.2 and should take 1–2 hr.

3. Incubate the culture at 42°C for 20 min with aeration.

4. Transfer the culture to 37°C and continue to shake or rotate. Cells should lyse in 1–2 hr.

5. Add 0.1 ml of chloroform and vortex.

6. Centrifuge at 4500g for 10 min to pellet the debris.

7. Carefully transfer the supernatant to a sterile, screw-capped tube. Add 0.1 ml of chloroform and vortex. Store the lysate at 4°C.

8. Spot-titer the lysate (see Appendix G, p. 239). Use L top agar containing 5 mM $CaCl_2$ and 10 mM $MgSO_4$ and L plates.

Notes

Do not overgrow the lysogens before induction, since this will result in poor phage yield.

Good phage induction is achieved by rapid temperature shifts. The use of shaking water baths is recommended. If the cells do not lyse in 3–4 hr, proceed with step 5. Although some cultures may not clear, they may still yield usable phage lysates.

PROCEDURE 12
Genetic Transduction Using P1*vir*

DAY 1

1. Prepare a culture tube with 5 ml of L medium.

2. Inoculate a single colony of each recipient strain and shake or rotate at 37°C overnight.

DAY 2

1. Centrifuge the overnight culture at $1500g$ for 10 min and resuspend the cell pellet in 2.5 ml of 10 mM $MgSO_4$ containing 5 mM $CaCl_2$.

2. To five small test tubes add recipient cells and P1*vir* grown on the donor strain as follows:

Test tube	Cells	P1 lysate
1	0.1 ml	—
2	0.1 ml	10 µl
3	0.1 ml	50 µl
4	0.1 ml	0.1 ml
5	—	0.1 ml

3. Incubate the five test tubes for 30 min at 30°C without shaking.

4. Add 0.1 ml of 1 M sodium citrate to each tube and mix. See Notes if phenotypic expression is required.

5. Add 2.5 ml of appropriate molten (45°C) top agar to each tube and plate on selective media. If the transduction is to be plated on indicator agar or if the transductants are to be replica plated, do not use top agar. If volumes to be spread are large (>0.3 ml), concentrate the cells by centrifugation and resuspend in 0.1 ml of L medium containing 20 mM citrate. Make sure plates are spread to dryness before incubation.

Notes

Sodium citrate is added to chelate the Ca^{++} and prevent reinfection of cells by P1*vir*.

Transductants should appear on plates prepared from tubes 2, 3, and 4. Tubes 1 and 5 control for reversion and lysate contamination, respectively.

The genetic marker being transduced may require phenotypic expression prior to plating the cells on selective media. For example, if the transduction is for a drug-resistance marker, 1 ml of L medium can be added after step 4, and the tubes can be incubated for 1 hr at the appropriate growth temperature without shaking prior to plating on selective media.

This procedure can also be employed for transduction with P1Tn*9clr*100. When using this phage, remember that it can form lysogens at 30°C. Consequently, all transductions should be done at 37°C.

PROCEDURE 13
Preparation of MudI(*lac*, Ap) Lysates

DAY 1

1. Prepare a culture tube with 5 ml of L medium.

2. Inoculate a single colony of a MudI(*lac*, Ap) lysogen (MAL103) and shake or rotate at 30°C overnight.

DAY 2

1. Inoculate 0.05 ml of the overnight culture in 10 ml of L medium.

2. Incubate at 30°C with aeration until the cells reach early-log phase. This corresponds to an OD_{600} of ~0.2 and should take 1–2 hr.

3. Incubate the culture at 42°C for 20 min.

4. Transfer the culture to 37°C and continue to shake or rotate. Cells should lyse in 1–2 hr.

5. Add 0.1 ml of chloroform and vortex to mix.

6. Centrifuge at 4500g for 10 min to pellet debris.

7. Carefully transfer the supernatant to a sterile, screw-capped tube. Add 0.1 of chloroform and vortex. Store the lysate at 4°C.

Notes

Do not overgrow the lysogens before induction since this will result in poor phage yield. Unlike λ or P1 lysates, Mu lysates become inactive during long-term storage at 4°C. It is wise to prepare a fresh lysate by heat induction for each new experiment.

PROCEDURE 14
Transduction with MudI(lac, Ap)

DAY 1

1. Prepare a culture tube with 5 ml of L medium.

2. Inoculate a single colony of the recipient strain and shake or rotate at 37°C overnight.

DAY 2

1. Centrifuge the overnight culture at 1500g for 10 min and resuspend the cell pellet in 2.5 ml of 10 mM $MgSO_4$ containing 5 mM $CaCl_2$.

2. Dilute the MudI(lac, Ap) lysate 10^{-2} and 10^{-3} in λ-dil.

3. Prepare four small test tubes as follows:

Test tube	Cells	MudI(lac, Ap)
1	0.1 ml	0.1 ml 10^{-2} dilution
2	0.1 ml	0.1 ml 10^{-3} dilution
3	0.1 ml	—
4	—	0.1 ml undiluted lysate

4. Mix the tubes and incubate for 20 min at 30°C without shaking.

5. Spread the sample on L plates or other selective media containing 25 µg/ml ampicillin.

6. Incubate the plates at 30°C.

Note

Ampicillin-resistant colonies should appear on the transduction plates (test tubes 1 and 2) and not on the control plates (test tubes 3 and 4).

PROCEDURE 15
Conversion of a MudI(lac, Ap) Lysogen to a λ Lysogen

DAY 1

1. Prepare a culture tube with 5 ml of L medium.

2. Inoculate a single colony of a MudI(lac, Ap) lysogen and shake or rotate at 30°C overnight.

DAY 2 (early in the day)

1. Centrifuge the overnight culture at 1500g for 10 min and resuspend the cell pellet in 2.5 ml of 10 mM $MgSO_4$.

2. Mix 0.1 ml of cells and 2.5 ml of molten (45°C) L top agar and pour on an L plate. Allow top agar to harden (5–10 min).

3. Spot the top agar lawn with three 50-μl aliquots of a λpSG1 lysate (~10^{10} phage/ml): undiluted, diluted 10^{-2}, and diluted 10^{-4}.

4. Incubate the plate at 30°C (do not invert).

DAY 2 (late in the day) **or DAY 3**

1. Spots of lysis should be evident on the lawn.

2. With a sterile toothpick, pick from the center the lysed area and streak on L plates containing 50 μg/ml chloramphenicol.

3. Incubate overnight at 30°C.

DAY 3 or DAY 4

1. Select large, healthy colonies as potential λ lysogens and restreak on lactose MacConkey plates containing 25 μg/ml chloramphenicol at 30°C to purify and check Lac phenotype.

Procedure 15

DAY 4 or DAY 5 (early in the day)

1. Test a single colony of each potential λ lysogen by cross-streaking against λcI and λvir at 30°C (see Notes below).

2. Streak the lysogens on L plates. Incubate overnight at 42°C to select cells that have cured the MudI(lac, Ap) prophage.

DAY 5 or DAY 6

1. Select colonies that grow at 42°C. Test resistance to chloramphenicol and ampicillin. Also test these survivors for $imm\lambda$. Save the colonies that are $imm\lambda$, Ap^s, Cm^s, and Lac^+.

Notes

This procedure is described in detail in Experiment 1 (p. 7).

Streak potential lysogens from the spot (Day 2 or 3) obtained with the highest phage dilution that shows confluent lysis. For a description of the cross-streak test for lysogens, see Procedure 6 (p. 99). For most experiments, cross-streak tests for immunity, if done early in the day, can be read the same day. However, always continue to incubate these tests overnight to confirm the results.

PROCEDURE 16

Conversion of a λcI⁺ Fusion Strain to a λcIts857 Lysogen

DAY 1

1. Prepare a culture tube with 5 ml of L medium.

2. Inoculate a single colony of a λcI⁺ fusion strain and shake or rotate at 37°C overnight.

DAY 2

1. Centrifuge the overnight culture at $1500g$ for 10 min and resuspend in 2.5 ml of 10 mM MgSO$_4$.

2. Add 0.1 ml of cells and 0.1 ml of λp1081.1 ($\sim 10^9$–10^{10} phage/ml) to a small test tube.

3. Adsorb phage at room temperature for 5 min.

4. Add 1.0 ml of L medium to the tube and incubate at 30°C for 1 hr.

5. Spread 3 L plates containing kanamycin (25 µg/ml) and Xgal with 0.1 ml of the mix: undiluted, diluted 10^{-1}, and diluted 10^{-2}.

6. Incubate the plates overnight at 30°C.

DAY 3

1. There should be three classes of Km^r transductant colonies: light blue, dark blue, and white. Pick light blue colonies and streak on L plates containing Xgal. For a discussion of these classes, see Experiment 10 (p. 63).

2. Incubate the plates overnight at 30°C.

DAY 4

1. Pick white colonies and restreak these onto two L plates.

2. Incubate one L plate at 30°C and the other plate at 42°C overnight.

DAY 5

1. Colonies that grow at 30°C and not at 42°C contain *imm*λ*cIts*857 and not *imm*λ*cI*$^+$.

2. Restreak all potential λ*cIts*857 lysogens on L plates and incubate at 30°C and 42°C to verify results.

Notes

The crossover events required for this immunity conversion are depicted in Figure 13 (p. 65).

The phage λ*p*1081.1 is an *att*$^+$ phage. However, this phage also carries an *int* amber (*int*am) mutation. Using this protocol in a Su$^-$ strain, about 70% lysogenize by homologous recombination at the existing prophage. The remaining 30% of the lysogens will have a prophage at the *att*λ site. These lysogens probably result from leakiness of the *int*am mutation. These can be verified by showing genetic linkage to *gal* by P1 transduction or by prophage induction and formation of *gal* or *bio* transducing phages. Of course, in a Su$^+$ (amber suppressor) strain, greater than 90% of the lysogens will have prophage at the *att*λ site. Another way of lowering the background of integration by site-specific recombination is to use a host strain that has been deleted for *att*λ.

PROCEDURE 17

Transfer of Tn*10* from λNK561 to the *Escherichia coli* Chromosome and Preparation of a Random Tn*10* Pool

DAY 1

1. Prepare a culture tube with 5 ml of L medium.

2. Inoculate a single colony of a P1Tn*9clr*100 lysogen and shake or rotate at 30°C overnight.

DAY 2

1. Centrifuge the overnight culture at 1500*g* for 10 min and resuspend the cell pellet in 5 ml of 10 mM $MgSO_4$.

2. Inoculate 20 ml of L medium with 0.05 ml of the $MgSO_4$ suspension.

3. Grow to mid-log phase ($OD_{600}=0.5$) at 30°C (1–3 hr).

4. Place 1 ml of cells into each of 15 small test tubes.

5. To each tube add 50 μl of a high-titer lysate ($>10^{10}$/ml) of λNK561 propagated on strain SG265.

6. Incubate for 5 min at room temperature for phage adsorption.

7. To each tube add 1 ml of L medium containing 40 mM sodium citrate and incubate with shaking for 1 hr at 30°C.

8. Centrifuge the entire culture at 1500*g* for 10 min.

9. Resuspend the cell pellet in 0.2 ml of L medium containing 20 mM sodium citrate.

10. Spread the entire sample on L citrate plates containing 25 μg/ml tetracycline.

11. Incubate the plates at 30°C overnight.

120   Procedure 17

DAY 3

1. Add 1 ml of L medium containing 20 mM sodium citrate to each plate.

2. Carefully loosen the colonies and wash into a centrifuge tube.

3. Rinse the surface of each plate again with 2 ml of L medium containing 20 mM sodium citrate and add this to the plate washings.

4. Centrifuge at 1500g for 10 min.

5. Resuspend the cell pellet in 10 ml of L medium containing 20 mM sodium citrate.

6. Repeat steps 4 and 5.

7. Inoculate 10 ml of L medium containing 25 µg/ml tetracycline with 0.01 ml of the cell suspension.

8. Shake at 30°C for 1.5–3 hr until cells reach early-log phase ($OD_{600} \simeq 0.2$). At this point, P1Tn9clr100 lysates can be made by thermal induction (see Procedure 11, p. 109). If you plan to store the pool, continue with step 9.

9. Centrifuge the entire culture at 1500g for 10 min.

10. Resuspend the cell pellet in 20 ml of L medium containing 10 mM $MgSO_4$.

11. Pooled cells can be frozen (see Appendix D, p. 231).

PROCEDURE 18

Isolation of Chromosomal Deletion Mutants Following λ Induction

DAY 1 (early in the day)

1. Streak strains containing a λcIts857 prophage on L plates.

2. Incubate at 30°C. Be sure to incubate the plates for at least 24 hr to get large colonies.

DAY 2

1. Using a sterile toothpick, scoop up single colonies and transfer to small test tubes containing 0.1 ml of L medium. Pick at least five colonies (one colony per tube).

2. Vortex the tubes to resuspend each colony.

3. Spread the contents of each tube on L plates (with supplements, if needed) prewarmed to 42°C. Work fast so the plates do not cool down.

4. Quickly return the plates to 42°C and incubate overnight.

DAY 3

1. Examine plates and pick no more than two temperature-resistant colonies from each plate.

2. Restreak each isolate again on L plates (with supplements, if needed) at 42°C.

DAY 4

1. Pick a single colony from each purified, temperature-resistant isolate and test for the desired deletions.

2. Test the isolates for prophage genes by marker rescue (see Procedure 9, p. 105).

Notes

Do not plate too many cells at 42°C (Day 2, step 3) or problems will be evident: There will be a large background, and many false temperature-resistant survivors will be isolated.

An alternative method to follow on Day 2 is to pick single colonies from the plate incubated at 30°C overnight and make small patches (1 patch per colony) on an L plate, prewarmed at 42°C. It is convenient to put about 40 patches on a single plate. On Day 3, pick one temperature-resistant survivor from each patch.

PROCEDURE 19
Selection of Deletion Mutants of λ by Using EDTA Plates

DAY 1 *(late in the day)*

1. Obtain single plaques by plating out dilutions of a phage stock on LE392, using TB plates (see Appendix G, p. 239).

2. Incubate the plates at 37°C overnight.

DAY 2 *(early in the day)*

1. Using sterile toothpicks, stab 20 single plaques to a fresh lawn of LE392.

2. Incubate the plates at 37°C for no more than 6–8 hr. Fresh stabs contain the most phage.

3. Harvest stabs by plugging the entire lysed area around the stab with a 1-ml disposable plastic pipette and by suspending the agar plug in 1 ml of TB with no Mg^{++}. Add a drop of chloroform. This suspension contains 10^8–10^9 phage and is called a pickate.

DAY 2 *(late in the day)*

1. Mix 1 µl and 10 µl of each pickate with 50 µl of a suspension of LE392 cells (*no* Mg^{++}) and plate both on TB-EDTA plates, using top agar that contains *no* Mg^{++}. Plates can be conserved by using 1 ml of molten (45°C) TB top agar containing no Mg^{++} and pouring two lawns per plate.

2. Incubate the plates at 37°C overnight.

DAY 3 *(early in the day)*

1. Pick an EDTA-resistant plaque from *each* pickate to 0.1 ml of TB (*no* Mg^{++}) and purify either by streaking or by plating 1 μl and 10 μl on LE392 cells (*no* Mg^{++}). Do either method on TB-EDTA plates, using top agar that contains *no* Mg^{++}.

2. Incubate the plates at 37°C for 6–8 hr.

DAY 3 *(late in the day)*

1. Once again, using sterile toothpicks, stab one plaque from each plate to a fresh LE392 lawn on normal TB plates (*no* EDTA). This step is to amplify each potential deletion for further testing. It is more reproducible than analysis of single plaques and more convenient than plate stocks.

2. Incubate the plates at 37°C for 6–8 hr.

DAY 4

1. Harvest each stab with a 1-ml disposable plastic pipette to 1 ml of TB containing 0.05 ml of chloroform. Test for genetic markers. Verify EDTA resistance. Verify plating on *recA* strains.

2. Make phage plate stocks (see Procedure 2, p. 91) of interesting mutants. Verify EDTA resistance and genetic markers of plate stocks.

Notes

Below is a list of λ phages for titrating the deletion size selection of EDTA plates. *Always include dilutions of the parent phage.* A valuable use of this set is to estimate the size of deletions in EDTA-resistant phages. An example of this titration is shown in Figure 19.

Phage	Percent of wild-type λ (48.5 kb)
Y1 (λ cIts857)	100
G6 (λ imm434 cI)	97
B10 (λ imm21 cI)	95
Y2 (λ b2 cIts857)	87.1

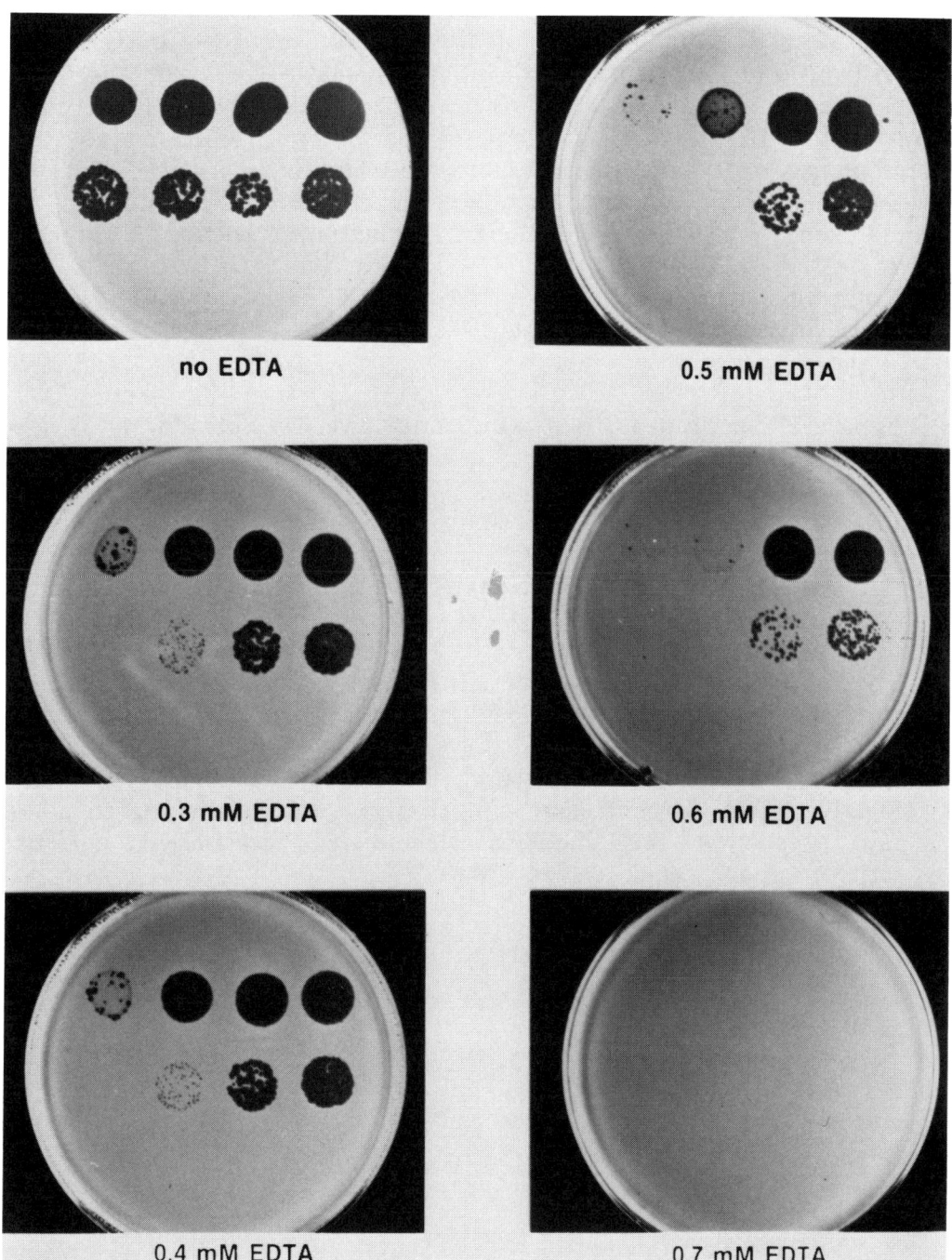

Figure 19
EDTA selection for deletion mutations in λ. This is a photograph of TB-EDTA plates containing different concentrations of EDTA prepared as described above. Lawns of a λ-sensitive strain in TB top agar (no Mg^{++}) were spotted with 10 µl of dilutions containing 10^5 and 10^2 phage. These phages from left to right are: Y1 (λ cIts857), G6 (λ imm434 cI), B10 (λ imm21 cI), and Y2 (λ b2 cIts857).

Normal phage stocks of 10^{10}/ml should be diluted 10^{-3} and 10^{-6} in TB without Mg^{++}. The Mg^{++} will inhibit the EDTA killing. Spot droplets (20–50 μl) from dilutions of the above phages on lawns of LE392 using TB plates and TB-EDTA plates. Use TB top agar again with *no* Mg^{++} and *no* additional EDTA. You should be able to spot all control phages plus your tester phages on a single plate. Let the droplets dry (with the petri dish cover partially off) and then incubate the covered plates at 37°C. You should see the killing effects of the EDTA maximized with λ and decreasing to λb2, which should be totally resistant.

The proper concentration of EDTA in TB-EDTA plates is determined empirically, using the phage deletion set as your tester phage. Prepare TB-EDTA plates as follows:

1. Prepare TB medium for plates and autoclave. Do not add $MgSO_4$ (see Appendix A, p. 217).

2. Add sterile 250 mM EDTA stock solution for desired final EDTA concentration directly to molten (45°C) agar. Prepare a small number of plates representing a range of final EDTA concentrations from 0.3 mM to 1 mM in 0.1 mM steps.

3. Test the plates with the phages in the set listed above to determine optimal EDTA concentration.

4. EDTA plates must be freshly prepared for reproducible results. The effective EDTA concentration can change dramatically within a few days. Accordingly, it is prudent to test plates immediately before use with the phage set described above.

PROCEDURE 20

Isolation of Deletion Mutations in λ Containing a Dam Allele

DAY 1 *(early in the day)*

1. Plate a 1-μl and 5-μl aliquot of a λDam lysate (titer of at least 10^{10}/ml on LE392) in top agar on TB plates using a non-amber-suppressing host (e.g., 594).

2. Incubate the plates at 37°C for 8 hr.

DAY 1 *(late in the day)*

1. Examine the plates. You should see about 500 plaques, 5–10% of which will be the same size as λ wild-type plaques and contain phage whose Dam mutation has reverted to wild type. The remainder of the plaques will be smaller, and the great majority of these plaques will be derived from phage containing deletions.

2. Pick a series of small plaques, each containing about 10^4 phage, and transfer to tubes, each with 1 ml of λ-dil and 0.05 ml of chloroform.

3. To confirm the presence of a deletion in phage lacking the D protein, purify the putative deletion phage once on strain 594 and test EDTA sensitivity as follows (this procedure is *only* for phage lacking the D protein):

    a. Spot a drop of a λ-dil plaque suspension on TB plates with agar overlays of either LE392 or 594. Dry the spots and incubate the plates at 37°C for 8 hr.

    b. Remove an agar plug from the center of each of the cleared spots and add to a tube containing 1 ml of λ-dil and a drop of chloroform.

    c. Dilute 20 μl of each of the two phage suspensions into tubes with either 200 μl of 20 mM Tris (pH 7.5), containing 20 mM EDTA, or 200 μl of λ-dil. Incubate on ice for 20 min.

d. Stop the EDTA reaction by adding 100 μl of 0.2 M MgSO$_4$ to each tube.

e. Assay each tube for phage by adding directly to each tube one drop of fresh overnight culture of LE392. Incubate at room temperature for 5 min and then add 3 ml of top agar and pour on TB plates.

f. Incubate the plates at 37°C overnight.

g. Dam phage carrying deletions are very sensitive to EDTA even at 4°C if they have been grown in an Su$^-$ strain (594) so that they lack the D protein. They become resistant to EDTA if they have been grown in an Su$^+$ strain (LE392) so that they contain the D protein. Thus, the titer of phage derived from plaques obtained from the 594 plate and treated with EDTA should be 100-fold to 1000-fold lower than the titer of the same phage treated with λ-dil. The titer of phage from the LE392 plate should be the same, regardless of the buffer in which the phage were incubated. Once the presence of a deletion has been confirmed, phage lysates are prepared using LE392.

Notes

Because of the increased sensitivity to inactivation of Dam-deleted phage grown in an Su$^-$ host like 594, the following two conditions must be met in order for this isolation procedure to be successful. First, the media (agar and broth) used to grow these phage must always contain 10 mM MgSO$_4$. Second, plaques are best picked into λ-dil after only 8 hr of growth on TB plates, rather than after overnight incubation, because the phage titer drops drastically after 8 hr.

The extent of the deletion and the location of deletion endpoints can be assessed by a variety of techniques, including CsCl gradients (see Procedure 33, p. 160), heteroduplex analysis, and marker-rescue experiments (see Procedure 9, p. 105).

PROCEDURE 21
Nitrosoguanidine Mutagenesis

DAY 1

1. Prepare a culture tube with 5 ml of L medium.

2. Inoculate a single colony of the strain to be mutagenized and shake or rotate at 30°C overnight.

DAY 2

1. Inoculate 0.1 ml of the overnight culture in 5 ml of L medium.

2. Shake or rotate at 37°C until the cells reach late-log phase. This corresponds to an OD_{600} of ~1.

3. To a centrifuge tube add 50 µl of nitrosoguanidine (2.5 mg/ml in 95% ethanol; see **Caution** below).

4. Place 2 ml of cells into the tube with nitrosoguanidine. Mix well.

5. Incubate for 10 min at 37°C.

6. Centrifuge at 1500g for 10 min to pellet cells.

7. Remove and discard the supernatant.

8. Resuspend the cells in 2 ml of M63 medium.

9. Repeat steps 6–8.

10. Resuspend the cells in 5 ml of L medium and grow overnight. This allows mutant phenotypes to be expressed. If phenotypic expression is not required, the mutagenized culture can be plated on selective media at this point.

DAY 3

1. On each of three selective plates, spread 0.1 ml of the overnight culture: undiluted, diluted 10^{-1}, and 10^{-2}.

2. Alternatively, if this culture is to be used for local mutagenesis, go to Procedure 10, Day 2, step 1 (p. 107).

Notes

The mutagenized culture can be frozen and stored for future mutant isolations (see Appendix D, p. 231).

Caution: Nitrosoguanidine is a potent carcinogen, and all glassware and solutions containing it should be handled very carefully. Always work in a fume hood and wear protective clothing. Dispose of supernatants by collecting them in a beaker filled with dilute HCl ($\sim$2 N or 1:5 dilution of concentrated HCl). The pH of this solution should be kept less than 1 (as checked by pH paper). Inactivation by acid treatment forms nitrous acid, a weak, nonvolatile mutagen. Inactivation in alkaline conditions, a widely used method, leads to the formation of diazomethane, which is a potent, volatile mutagen. Therefore, decontaminate glassware exposed to nitrosoguanidine by treatment with acid.

PROCEDURE 22

Mutagenesis of λ by Ultraviolet Light

DAY 1

1. Prepare a culture tube with 5 ml of TB.

2. Inoculate a single colony of the appropriate host (e.g., 594 or MC4100) and shake or rotate at 30°C overnight.

3. Titer the stock of phages to be mutagenized (see Appendix G, p. 239).

DAY 2

1. Cells for this experiment must be in early- to mid-log phase. Inoculate 5 ml of TB with an appropriate dilution of the overnight cells so that they will be ready at step 4.

2. Calculate the titer of the phage stock and dilute in 5 ml of λ-dil to 1×10^9/ml.

3. In a plastic petri dish, with the lid *off*, expose 2 ml of the lysate, spread out as a thin film, to about 4000 ergs/mm^2 UV light. This treatment will reduce viable phage in the lysate about five orders of magnitude. The irradiated phage stock can be stored at 4°C and treated as a normal phage lysate.

4. Perform the following operations in dim light:

   a. Mix 0.1 ml of log-phase cells and 0.1 ml of the irradiated phage lysate as a droplet in a plastic petri dish. Wait 5 min at room temperature.

   b. Expose the droplet to 300 ergs/mm^2 UV light.

   c. Add 10-μl samples to amber or foil-covered culture tubes with 1 ml of L medium containing 10 mM MgSO$_4$. Prepare about 20 tubes.

5. Incubate at 37°C with shaking for 2 hr.

Procedure 22

6. Add 0.05 ml of chloroform per tube. Examine the lysate for mutants. For example, if you are mutagenizing a λ*plac* fusion phage, screen the Lac phenotype on TB plates containing Xgal (see Procedure 8, p. 104). For an indication of mutagenesis, look for clear plaques if you are using a turbid phage. Good mutagenesis is indicated by 1–2% clear plaques.

DAY 3

1. Examine the plates. Pick mutant plaques to lawns of appropriate bacteria or to 0.1 ml of TB containing 50 µl of chloroform.

2. Purify by streaking or replating 1 µl and 10 µl of plaque suspension on the appropriate bacterial strain and retest for the mutant phenotype.

DAY 4

1. Examine the plates and pick one purified mutant plaque. Prepare plate stocks or liquid lysates of each mutant phage (see Procedures 1, 2; pp. 89, 91).

2. Verify the mutant phenotype.

Notes

The instructions for Day 2, step 4c, recommend individual 1-ml phage lysates. This is to avoid picking identical mutant phage plaques on Day 3.

When mutagenizing chromosomal DNA carried on λ transducing phages, beware of recombination with the chromosomal homolog and subsequent rescue of the wild-type allele during propagation of the mutant phage. The best course of action is to use a chromosomal deletion of the appropriate gene(s) or to use a *recA* strain.

PROCEDURE 23
Hydroxylamine Mutagenesis of Phage

DAY 1

1. In a test tube, mix:

   a. 0.4 ml of phosphate-EDTA buffer (0.5 M potassium phosphate, pH 6.0, 5 mM EDTA);

   b. 0.5 ml of sterile water;

   c. 0.8 ml of hydroxylamine solution (1 M NH_2OH at pH 6.0, made fresh by adding 0.56 ml of 4 M NaOH to 0.35 g of NH_2OH and then adding sterile water to 5 ml);

   d. 20 µl of sterile 1 M $MgSO_4$.

2. Add 0.2 ml of phage stock (10^9–10^{11}/ml).

3. Include a control, using sterile water instead of hydroxylamine.

4. Incubate these mixtures at 37°C for 12–48 hr.

DAYS 2–3

1. Take 0.1-ml samples periodically (every 6–8 hr) into 10 ml of cold L medium containing 1 M NaCl and 1 mM EDTA. After 1 hr, dilute twofold in λ-dil and store at 4°C.

2. Plate some dilutions that will give 1000–3000 plaques. Examine the lysate for mutants. For example, if you are mutagenizing a λplac fusion phage, screen the Lac phenotype on TB plates containing Xgal (see Procedure 8, p. 104). For an indication of mutagenesis, look for clear plaques if you are using a turbid phage. Good mutagenesis is indicated by 1–2% clear plaques. Expect a few percent survival per 24 hr of incubation in hydroxylamine and 50–100% survival in control. Save dilutions at 4°C for large-scale plating later since they are more stable than the dilutions made in step 1.

PROCEDURE 24
mutD Mutagenesis of λ

DAY 1

1. Streak a *mutD* strain (LE30) on a minimal glucose plate. This media contains no thymine; consequently, mutator activity is reduced considerably (see Experiment 12, p. 75).

2. Incubate the plate at 37°C overnight.

DAY 2 *(early in the day)*

1. Pick five colonies and inoculate each into five culture tubes containing 5 ml of L medium. This medium contains sufficient thymine for optimal mutator activity.

2. Shake or rotate at 37°C until turbid (about 6–8 hr).

DAY 2 *(late in the day)*

1. Centrifuge the LE30 cultures at 1500g for 10 min. Resuspend each cell pellet in 2.5 ml of 10 mM $MgSO_4$.

2. Store the resuspended cells at 4°C.

3. Test each for mutator activity and sensitivity to λ by using λ*vir* in a cross-streak test (see Procedure 6, p. 99).

4. Incubate test plate at 37°C overnight.

DAY 3

1. Examine the cross-streak test plate. Use one LE30 culture that is λ sensitive yet exhibits significant λ-resistant survivors in the streak as compared with controls. It is not uncommon to find three out of five colonies tested to appear essentially λ resistant. These are often mucoid and should be discarded.

2. Make a plate stock (Procedure 2, p. 91) of the phage to be mutagenized, using the LE30 culture selected in step 1 above. Use 0.1 ml of the LE30 cells (Day 2, step 2) and 0.1 ml of a lysate with a titer of 1×10^6/ml.

DAY 4

1. Examine the lysate for mutants. For example, if you are mutagenizing a λ*plac* fusion phage, screen the Lac phenotype on TB plates containing Xgal (see Procedure 8, p. 104). For an indication of mutagenesis, look for clear plaques if you are using a turbid phage. Good mutagenesis is indicated by 1–2% clear plaques. Plate 3 shows results of a *mutD*-mutagenesis experiment screening for λ*int* and λ*xis* mutants, using the red-plaque test (Procedure 5, p. 97).

2. Pick mutant phages to appropriate bacterial lawns with toothpicks or to 0.1 ml of TB containing 50 µl of chloroform. Purify these presumptive mutants either by streaking or plating 1 µl and 10 µl of plaque suspension. Purify isolates at least once.

DAYS 5–6

1. Make plate stocks or liquid lysates of purified mutant phages.

2. Titer lysates and verify mutant phenotypes.

Notes

LE30 is Su$^-$, so it cannot be used to propagate phages with nonsense mutations in essential genes.

When mutagenizing chromosomal DNA carried on λ transducing phages, beware of recombination with the chromosomal homolog and subsequent rescue of the wild-type allele during propagation of the mutant phages. The best course of action is to use a chromosomal deletion of the appropriate gene(s) or a *recA* strain.

Unlike chemically mutagenized stocks, phages mutagenized by *mutD* propagation are as stable as wild-type phage lysates. If kept at 4°C in the dark and away from organic solvents and detergents, they should be stable for years. One *mutD* plate stock will be a long-time source of independent mutants.

mutD is a conditional mutator (see Experiment 12). Therefore, overnight L or TB cultures of LE30 are heavily mutagenized. We often observe reduced plating efficiencies of λ stocks on such lawns. This problem can be overcome by growing overnight cultures in minimal media, where mutator activity is substantially decreased.

PROCEDURE 25
DNA Extraction from Bacterial Cells

DAY 1

1. Prepare a 500-ml flask containing 100 ml of TB.

2. Inoculate with a single colony and shake at 30°C overnight.

DAY 2

1. Collect the cells by centrifugation at 2000g for 20 min.

2. Resuspend the cell pellet in 5 ml of 50 mM Tris·HCl, pH 8.0, 50 mM EDTA. Freeze the cell suspension at −20°C.

3. Make a fresh lysozyme solution, 10 mg/ml in 0.25 M Tris·HCl, pH 8.0. Add 0.5 ml of lysozyme solution to frozen cells and thaw with mixing in a room-temperature water bath. When just thawed, put on ice for 45 min.

4. Add 1 ml of STEP solution (see Notes below) and mix well. Heat at 50°C for 60 min with occasional, gentle mixing.

5. Add 6 ml of Tris-buffered phenol (see Procedure 40, p. 177). Mix gently for 5 min to emulsify. Do not vortex.

6. Centrifuge at 1000g for 15 min to separate the layers. DNA is in the top, aqueous layer.

7. Transfer the aqueous layer to a clean tube. Try not to include material at the interface. If the aqueous layer is contaminated with interface material, redo steps 5 and 6.

8. Add 0.1 volume of 3 M sodium acetate. Mix gently. Do not vortex.

9. Add 2 volumes of ethanol and invert to mix. DNA and RNA should precipitate as a glob.

Procedure 25

10. Spool out the precipitate with a glass micropipette. Remove excess alcohol by gently rotating the precipitate against the side of the tube. Transfer to a clean tube containing 5 ml of 50 mM Tris·HCl, pH 7.5, 1 mM EDTA, 200 µg/ml RNase A (Appendix A, p. 217).

11. To dissolve the precipitate, rock the tube gently at 4°C overnight.

DAY 3

1. It is essential to have the DNA completely dissolved before continuing. If necessary, add more buffer.

2. Add an equal volume of chloroform and invert several times to mix. Try to emulsify. Do not vortex.

3. Centrifuge at 1000g for 15 min to separate layers.

4. Transfer the aqueous top layer to a clean tube. Add 0.1 volume of 3 M sodium acetate, followed by 2 volumes of ethanol, and mix by inverting. DNA should precipitate as long threads.

5. Spool out the DNA with a glass micropipette as described in step 10. Dissolve *completely* in 2 ml of 50 mM Tris·HCl, pH 7.5, 1 mM EDTA. More buffer may be necessary for high concentrations of DNA. Store at 4°C. Determine the concentration of DNA as described below.

Determination of DNA Concentration, Using Ethidium Bromide – Agarose Plates

1. Prepare ethidium bromide–agarose plates: Boil to dissolve 1% agarose in distilled water. Add ethidium bromide to a final concentration of 5 µg/ml. Pour 20 ml into standard plastic petri dishes. Store solidified plates wrapped in aluminum foil at 4°C. Plates are stable for several weeks. Freshly poured plates do not work well because the DNA does not diffuse well in the agarose. Placing a freshly poured plate uncovered in a 37°C incubator for 30 min to dry it slightly helps to overcome this problem.

2. Prepare DNA concentration standards ranging from 1 µg/ml to 30 µg/ml in 5 µg/ml increments. Salmon sperm DNA is one example of DNA that can be used.

3. To determine the concentration of a particular DNA sample:

   a. Spot 5 µl of each standard on the agarose surface.

   b. Make appropriate dilutions of the unknown DNA sample and spot 5 µl of each dilution onto the agarose surface.

   c. Put the plate into a 37°C incubator to allow the liquid in the spots to absorb into the plate and to allow the DNA to diffuse (10–15 min).

   d. Place the uncovered dish on a UV illuminator to visualize the ethidium bromide–stained DNA. Compare the intensity of the ethidium bromide–stained DNA with that of the standards.

Notes

This procedure has been used to extract DNA from a wide variety of enteric bacteria. It yields partially purified DNA of sufficient quality to permit restriction enzyme digestion, ligation, and cloning. This protocol can be scaled down by a factor of 10.

Some strains are more difficult to lyse than others. For the recalcitrant strains, increase the incubation at 50°C in step 4 of Day 2 to 60 min.

You should expect to obtain 0.5–1.0 mg of DNA from 100 ml of culture by using this procedure.

Aliquots of DNA may be conveniently dialyzed using the drop-dialysis method (see Procedure 42, p. 182).

STEP is *S*DS, *T*ris, *E*DTA, and *p*roteinase K. This solution is made as follows:

   0.5% SDS
   50 mM Tris·HCl, pH 7.5
   0.4 M EDTA

   Store at room temperature.

Add proteinase K powder to a final concentration of 1 mg/ml immediately before use. Store proteinase K powder at 4°C.

PROCEDURE 26
Large-scale Isolation of λ DNA

1. Start with a λ lysate prepared as described in Procedure 3 (p. 93).

2. Centrifuge at 6000g for 10 min. Decant the supernatant into a beaker or flask, trying not to get any of the chloroform. Save 1 ml to determine the phage titer.

3. Add polyethylene glycol (PEG) powder to 10% w:v (10 g PEG 6000/100 ml lysate). Shake or stir at room temperature to dissolve completely.

4. Put the beaker or flask in ice water for 60 min. The phage will coprecipitate with PEG.

5. Collect the pellet by centrifugation at 6000g for 10 min at 4°C.

6. Resuspend the pellet in 50 mM Tris·HCl, pH 7.5, 10 mM $MgSO_4$ (TM buffer). Use a minimum amount of TM: 5–10 ml per liter of lysate. Use an antifoaming agent to eliminate bubbles (50 μl).

7. Extract the PEG-phage suspension with an equal volume of chloroform in a screw-capped container by gently inverting the tube for 1 min at room temperature.

8. Centrifuge at 2000g for 10 min. The chloroform is at the bottom; the PEG is the white interface. Pipette the aqueous, upper layer containing the phage into a sterile tube. Sometimes significant amounts of phage can be trapped at the interface. They can be recovered by reextracting the PEG-chloroform and interface with an equal volume of fresh TM.

9. Make a step gradient in an ultracentrifuge tube (see Procedure 4, p. 95). Add the aqueous layer from step 8. Fill to the top with TM buffer.

10. Centrifuge at 35,000 rpm for 60 min at 4°C.

11. Decant the supernatant and save the phage pellet.

12. Resuspend the pellet in 1 ml of TM buffer, using a pasteur pipette and bulb. Add RNase A and DNase I to a final concentration of 10 μg/ml and 1 μg/ml, respectively (see Appendix A, p. 217). Incubate for 30 min at 37°C.

13. Add 0.2 final volume of STEP buffer (for preparation, see Procedure 25, p. 137). Mix well.

14. Heat at 50°C for 15 min. Transfer to a sterile polypropylene tube.

15. Add an equal volume of Tris-saturated phenol and extract by rocking the test tube (see Procedure 40, p. 177).

16. Centrifuge at 1000g for 10 min. Remove the phenol phase (leave the interphase) and reextract the aqueous phase with phenol:chloroform:isoamyl alcohol (25:24:1).

17. Centrifuge at 1000g for 5 min. Remove the top, aqueous layer with a pasteur pipette and transfer to a sterile polypropylene tube. Extract with an equal volume of chloroform:isoamyl alcohol (24:1).

18. Centrifuge at 1000g for 5 min. Remove the top, aqueous layer with a pasteur pipette and transfer to a sterile polypropylene tube.

19. This DNA solution can be dialyzed against the proper buffer (see Procedure 42, p. 182), or 2 volumes of 95% ethanol can be added to precipitate the DNA. It is best to spool out DNA on a glass micropipette rather than pellet in a centrifuge. Significant amounts of SDS will be present and will coprecipitate with the DNA if more than 2 volumes of ethanol are added. If SDS and DNA coprecipitate, the DNA will not be digested efficiently with most restriction enzymes.

20. The DNA concentration can be readily determined by spotting dilutions on agarose–ethidium bromide plates (see Procedure 25).

Note

Do not increase the time at 0°C in step 4 or debris will begin to precipitate as well as phage.

PROCEDURE 27
Rapid Isolation of λ DNA

1. Put 0.6 ml of a liquid lysate (see Procedure 1, p. 89) in a microfuge tube. Phage titer must be at least 10^{10}/ml.

2. Add 0.6 ml of a DEAE-cellulose slurry equilibrated with L medium (see below).

3. Shake gently back and forth 20–30 times.

4. Remove the DEAE-cellulose by centrifugation in a microfuge at ~12,000g for 5 min and transfer the supernatant to a new microfuge tube.

5. Repeat step 4 to remove any remaining DEAE-cellulose.

6. To 0.8 ml of supernatant, add 100 μl of 5 M NaCl.

7. Add 540 μl of isopropanol and chill for 20 min at −20°C.

8. Centrifuge for 5 min in a microfuge. Pour off and discard the supernatant.

9. Add 1 ml of 70% ethanol. Invert several times to mix.

10. Centrifuge for 5 min in a microfuge to collect the precipitate. Pour off the supernatant. Resuspend the pellet in 200 μl of TE (10 mM Tris·HCl, 1 mM EDTA, pH 8.0).

11. Add 200 μl of phenol. Vortex briefly. Centrifuge for 5 min in a microfuge. Remove the top, aqueous layer containing DNA to a clean microfuge tube.

12. Repeat step 11.

13. Precipitate the DNA from the aqueous layer by adding 15 µl of 5 M NaCl and 2 volumes of 95% ethanol (see Procedure 41, p. 180). Resuspend the dry ethanol precipitate in 100 µl of TE.

14. Use 30–50 µl for each restriction enzyme reaction. DNA may be further purified by drop dialysis (see Procedure 42, p. 182).

Preparation of DEAE-Cellulose

1. Place the DEAE-cellulose (DE52, Whatman) into several volumes of 0.05 N HCl (check that the pH is below 4.5).

2. With stirring, add concentrated NaOH until the pH approaches that of L medium (pH 6.8).

3. Let the resin settle and decant the supernatant. Add several volumes of L medium, mix, allow to settle, and decant.

4. Repeat L-medium rinses in step 3 until the pH and conductivity of the supernatant are identical to L medium (pH 6.8).

5. Resuspend in a final slurry of about 75% resin and 25% L medium. Add sodium azide (0.1%) to prevent contamination.

Notes

Phages from plate stocks yield DNA that is not cut well by various enzymes.

This procedure was developed by R. Taylor and S. Benson (National Cancer Institute, Frederick Cancer Research Facility) and M. Graham and M. Olson (Washington University Medical School).

PROCEDURE 28
Large-scale Isolation of Plasmid DNA

DAY 1

1. Prepare a culture tube with 10 ml of L medium containing the appropriate antibiotics.

2. Inoculate a single colony and shake or rotate at 37°C overnight.

DAY 2

1. Add 10 ml of overnight culture to 1 liter of L medium with the appropriate antibiotic in a 4-liter flask.

2. Shake at 37°C until the $OD_{600}=0.8-1.0$.

3. Amplify the plasmid copy number by adding 300 μg/ml spectinomycin or 150 μg/ml chloramphenicol and shaking the culture for at least 16 hr at 37°C.

DAY 3

1. Centrifuge the cells at 4500g for 15 min at 4°C.

2. Resuspend the cells in 30 ml of 25% sucrose in 50 mM Tris·HCl, pH 8.0.

3. Add 10 ml of lysozyme (5 mg/ml in 0.25 M Tris·HCl, pH 8.0) and 10 ml of EDTA (0.25 M, pH 8.0). Mix well.

4. Incubate for 15 min on ice.

5. Add 50 ml of the following detergent mix to the lysozyme-treated cells:

   5 ml 10% Triton X-100 in 0.25 M Tris·HCl, pH 8.0
   125 ml 0.25 M EDTA
   25 ml 1.0 M Tris·HCl, pH 8.0
   H_2O up to 500 ml

6. Incubate on ice for 30 min.

7. Centrifuge at 27,000 rpm in a Beckman SW27 or an equivalent rotor for 1.5 hr at 15°C. Save the supernatant; it contains the plasmid DNA. Chromosomal DNA remains in the pellet.

8. To the supernatant add predigested Pronase (see Notes below) to a final concentration of 250 µg/ml.

9. Incubate at 37°C for 30 min.

10. Extract by adding 0.5 volume of phenol saturated with 50 mM Tris, pH 8.0 (see Procedure 40, p. 177). Mix well and centrifuge at 1000g for 10 min at 4°C.

11. Remove and measure the volume of the top, aqueous layer. Precipitate the nucleic acids by adding 0.1 volume of 3 M sodium acetate and 2 volumes of ice-cold 95% ethanol. Mix well.

12. Chill at −20°C overnight or at −70°C for 30 min.

13. Collect the nucleic acid precipitate by centrifugation at 6000g for 20 min at 4°C.

14. Pour off the ethanol supernatant and remove the excess liquid with sterile absorbent paper.

15. Resuspend the dry nucleic acid pellet in 18 ml of TE-10 (20 mM Tris·HCl, 10 mM EDTA, pH 8.0).

16. Add 16 g of CsCl and 1 mg of ethidium bromide. The final density of this solution should be 1.58−1.59 g/ml.

17. Centrifuge at 65,000 rpm in a Beckman 70Ti rotor at 15°C for at least 16 hr. Use clear ultracentrifuge tubes that can be punctured with a needle. Excellent results can be obtained by using the Beckman 50Ti rotor at 40,000 rpm for 48 hr at 15°C.

DAY 5

1. Remove the centrifuge tubes carefully and examine them, using UV light. You should see two orange bands in the center of the tube. The top one is primarily chromosomal DNA and the bottom one is covalently closed, circular plasmid DNA. Hopefully, there will be much more of the bottom band because most of the chromosomal DNA was removed in step 7 of Day 3.

2. Use an 18-gauge needle to puncture the tube from the side to remove the plasmid band.

3. Add an equal volume of water-saturated isobutanol and mix well. This will extract the ethidium bromide. Remove the aqueous layer and extract at least three more times to remove all the red color (ethidium bromide).

4. Precipitate the plasmid DNA with ethanol as described in Procedure 41 (p. 180).

5. Resuspend the dry plasmid DNA pellet in 1 ml of TE (10 mM Tris·HCl, 1 mM EDTA, pH 8.0)

Notes

The Pronase stock added at step 8 of Day 3 is incubated for 30 min at 37°C (i.e., "predigested"). This incubation destroys contaminating enzyme activities. Pronase is resistant to autocatalyzed proteolysis.

The concentration of DNA can readily be determined by spotting dilutions on agarose–ethidium bromide plates (see Procedure 25, p. 137).

PROCEDURE 29
Methods for Rapid Plasmid DNA Isolation

METHOD A: *Rapid Plasmid Preparation by Alkaline Extraction*

Reagents

Lysozyme solution

    50 mM glucose
    10 mM EDTA
    25 mM Tris·HCl, pH 8.0

    Store at 4°C. Immediately prior to use, add lysozyme to 5 mg/ml.

NaOH-SDS stock

    0.2 N NaOH
    1% SDS

    Prepare weekly. Store at room temperature.

Potassium acetate stock

    60 ml 5 M potassium acetate
    28.5 ml glacial acetic acid
    11.5 ml H_2O
    (pH 4.8)

    Store at room temperature.

DAY 1

1. Prepare a culture tube with 5 ml of L medium containing the appropriate antibiotics.

2. Inoculate a single colony and shake or rotate at 37°C overnight.

DAY 2

1. Centrifuge the overnight culture at 1500g for 10 min.

2. Resuspend the cell pellet in 0.2 ml of lysozyme solution. Transfer to a 1.5-ml microfuge tube and incubate for 5 min on ice.

3. Add 0.4 ml of NaOH-SDS stock. Vortex gently; the solution should turn translucent. Incubate for 5 min on ice.

4. Add 0.3 ml of the potassium acetate stock and vortex gently. A precipitate should form. Freeze in a dry ice–ethanol bath for 5 min. Let thaw at room temperature.

5. Centrifuge for 15 min in a microfuge (~12,000g) to collect the precipitate.

6. Carefully remove 0.75 ml of the supernatant and transfer to a clean microfuge tube. Be careful not to take any material at the interface. Add 0.45 ml of isopropanol and mix well.

7. Place in a dry ice–ethanol bath for 5 min. Warm to room temperature and centrifuge for 5 min in a microfuge to pellet the DNA.

8. Remove the supernatant and rinse the pellet with 2 ml of cold 70% ethanol.

9. Centrifuge for 2 min in a microfuge. Remove and discard the ethanol rinse.

10. Dry the DNA pellet under vacuum. Add 0.1 ml of TE (10 mM Tris·HCl, 1 mM EDTA, pH 8.0) and vortex to resuspend the pellet.

11. Use 2 μl of the DNA for restriction enzyme analysis. Small volumes may be dialyzed, if necessary, by drop dialysis (see Procedure 42, p. 182).

Notes

This protocol can be scaled up such that 1 liter of cells can be used starting with 20 ml of lysozyme solution and proportionate amounts of other reagents.

After step 9, there is substantial cellular RNA present. If the digested DNA is to be displayed on low-percentage agarose gels, there is no need to treat the sample further. However, if the DNA is to be displayed on acrylamide gels, the RNA must be further digested to reduce the size of fragments that would comigrate with low-molecular-weight DNA fragments. This can be done by treating the DNA in 100 μl of TE with 1 μl of RNase A (see Appendix A, p. 217) for 15 min at room temperature. An alternative is to treat individual enzyme digests for 5 min before loading on a polyacrylamide gel.

This procedure is a modification of a published protocol (Birnboim and Doly 1979).

METHOD B: *Rapid Plasmid DNA Isolation by Phenol Extraction*

Reagents

NTE

>100 mM NaCl
>20 mM Tris·HCl, pH 8.0
>10 mM EDTA

>Store at room temperature.

Lysozyme solution

>250 mM Tris·HCl, pH 8.0
>5 mg/ml lysozyme

>Prepare immediately prior to use.

DAY 1

1. Prepare a culture tube with 5 ml of L medium containing the appropriate antibiotics.

2. Inoculate a single colony and shake or rotate at 37°C overnight.

DAY 2

1. Centrifuge 1.5 ml of the cells for 2 min in a microfuge.

2. Resuspend the cell pellet in 0.15 ml of NTE.

3. Add 15 µl of lysozyme solution. Incubate for 10 min at room temperature.

4. Extract gently for 3–5 min with an equal volume of phenol:chloroform:isoamyl alcohol mixture (25:24:1).

5. Centrifuge for 5 min in the microfuge.

6. Transfer 0.15 ml of the top, aqueous layer to a clean microfuge tube.

7. Extract once with an equal volume of chloroform:isoamyl alcohol (24:1).

8. Centrifuge for 1 min in the microfuge.

9. Remove the top, aqueous layer, add 2 volumes of ethanol, and chill in a dry ice–ethanol bath for 15 min.

10. Centrifuge for 15 min in the microfuge to collect the precipitate. Small volumes may be dialyzed, if necessary, by drop dialysis (see Procedure 42, p. 182).

11. Remove the supernatant and dry the pellet under vacuum.

12. Resuspend the pellet in 50 µl of TE (10 mM Tris·HCl, 1 mM EDTA, pH 8.0).

13. Use 10 µl of the DNA solution for restriction enzyme analysis.

Note

See Notes for Method A regarding cellular RNA in the final DNA preparation.

METHOD C: *Rapid Plasmid DNA Isolation by Boiling*

Reagents

STET

    8% Sucrose
    5% Triton X-100
    50 mM EDTA
    50 mM Tris·HCl, pH 8.0

DAY 1

1. Streak the plasmid-harboring strain on an appropriate antibiotic plate.

DAY 2

1. Pick up a broad swipe of cells with the side of a flat toothpick and resuspend the cells in 50 µl of cold STET in a microfuge tube by twirling the toothpick and then vortexing. Keep on ice until step 3.

2. Add 4 µl of a fresh 10 mg/ml lysozyme solution made in water.

3. Suspend the tube in a boiling water bath for 40–60 sec.

4. Centrifuge in a microfuge for 10 min at room temperature.

5. Remove the gelatinous pellet with a flat toothpick. Save the supernatant.

6. Add 40 µl of isopropanol to the supernatant in the original tube.

7. Chill for 5 min at −20°C or in a dry ice–ethanol bath.

8. Centrifuge for 5 min in a microfuge to collect the precipitate.

9. Remove the supernatant and dry the pellet under vacuum.

10. Resuspend the pellet in 15 µl of water.

Notes

This method requires the least amount of work and yields plasmid DNA that is readily digested by most enzymes. There is usually enough DNA for three enzyme digests.

For a good DNA yield, it is important that the tubes and the STET be cold before adding the cells. This method can be scaled up to accommodate 1 liter of culture (grown overnight without amplification). Resuspend this amount of cells in 34 ml of STET (Day 2, step 1); use 1 ml of lysozyme (Day 2, step 2). Carefully heat until the cell suspension begins to boil. Transfer to a boiling water bath (Day 2, step 3). Centrifuge at 45,000g for 15 min to pellet debris. Carefully remove 18 ml of supernatant and prepare CsCl gradients as described in Day 3 of Procedure 28 (p. 144).

It is also possible to isolate sufficient plasmid DNA from single colonies for analysis. Plasmids isolated directly from selected transformants can be screened for alterations in size by agarose gel electrophoresis of half of the supernatant obtained at step 5. Although convenient, this method yields plasmid DNA contaminated with variable amounts of chromosomal DNA.

This procedure is a modification of a published protocol (Holmes and Quigley 1981).

PROCEDURE 30
λ Hybrid Formation

I. RESTRICTION ENZYME DIGESTION

1. Before beginning the digestion, clean up your DNA by drop dialysis (see Procedure 42, p. 182). Digest vector and insert DNAs in separate tubes. When choosing DNA concentrations, remember that the final ligation reaction should contain, in 10 μl, 1–2 μg of vector and 0.2–6.0 μg of insert. It is often useful to use three different concentrations of the insert DNA: 0.2–0.5 μg; 1–2 μg, and 4–6 μg. In this case the vector DNA is always kept constant at 1–2 μg. In the following example, only one concentration of insert DNA is used. In 1.5-ml microfuge tubes, set up the following two reactions:

	Vector reaction	Insert reaction
H$_2$O	X	X
10× Buffer	5 μl	5 μl
DNA	X	X
Enzyme	X	X
Total volume	50 μl	50 μl

   The amounts of H$_2$O, DNA, and enzyme will vary with DNA concentration and specific activity of the restriction enzyme.

2. Incubate digests for 60 min at 37°C. Verify complete digestion by running an aliquot on an agarose gel. If digestion is not complete, continue incubation at 37°C.

3. Heat at 70°C for 3 min to inactivate the restriction enzyme.

4. Add an equal volume of chloroform and mix by vortexing.

5. Centrifuge for 1 min in a microfuge (~12,000g).

6. Remove and *combine* the top, aqueous layers from the vector and insert reactions in a clean microfuge tube.

7. Add 0.1 volume of 3 M sodium acetate to the combined supernatants.

8. Add 2.5 volumes of ethanol; mix and chill in a dry ice–ethanol bath for 5 min. (The procedure can be stopped at this point by storing the ethanol precipitation at −70°C.)

9. Centrifuge for 5 min in a microfuge in the cold room.

10. Carefully decant and discard the supernatant; dry the pellet under vacuum.

II. LIGATION REACTION

1. Resuspend the pellet from step 10 above in 10 µl of 1× ligase buffer (see Notes below). Remove 1–2 µl for preligation control and freeze at −20°C. Keep the remainder on ice.

2. Add 1 µl of T4 ligase (5–10 units) and mix gently.

3. Incubate at 4°C for 6–18 hr.

4. Withdraw 1–2 µl, mix with loading solution, and check the ligation by running both the preligation and postligation samples on a 0.7% agarose gel (see Procedure 43, p. 183).

5. Transfect (see Procedure 38, p. 171) or package DNA (see Procedure 39, p. 173).

Notes

Stock 10× ligase buffer (−ATP)

    660 mM Tris·HCl, pH 7.5
    66 mM MgCl$_2$
    100 mM dithiothreitol (DTT)

Add 10 µl of 0.1 M ATP to 90 µl of 10× ligase buffer immediately before use. Dilute 1:10 for 1× ligase buffer.

In preparing the 0.1 M ATP stock, be sure to neutralize to pH 7 with KOH. Store samples of this stock at −20°C. If you use in vitro packaging, it is important to extract the ligation mixture with an equal volume of chloroform–isoamyl alcohol (24:1), followed by ethanol precipitation (see Procedures 40, 41, pp. 177, 180). Dissolve the pellet in 10–20 µl of TE (10 mM Tris·HCl, 1 mM EDTA, pH 8.0) and then use in vitro packaging. The chloroform–isoamyl alcohol extraction removes potential inhibitors of in vitro packaging.

For a detailed discussion of ligation theory, see Maniatis et al. (1982).

PROCEDURE 31
Plasmid Hybrid Formation

DAY 1

1. Before beginning the digestion, clean up your DNA by drop dialysis (see Procedure 42, p. 182). Digest the plasmid vector and insert DNAs with the appropriate restriction enzymes in separate reactions. Use 20–50 µl of reaction volume as described in Procedure 30 (p. 152). Use equimolar amounts of plasmid and insert DNAs. In most experiments the total amount of DNA ranges between 0.1 µg and 1.0 µg. Incubate the digests for 60 min at 37°C.

2. Heat at 70°C for 3 min to inactivate the restriction enzymes.

3. Add an equal volume of chloroform and mix by vortexing.

4. Centrifuge for 1 min in the microfuge (~12,000g).

5. Remove and combine the top, aqueous layers from the vector and insert reactions in a clean microfuge tube.

6. Add 0.1 volume of 3 M sodium acetate to the combined supernatants.

7. Add 2.5 volumes of ethanol; mix and chill in a dry ice–ethanol bath for 5 min.

8. Centrifuge for 5 min in a microfuge.

9. Carefully decant and discard the supernatant; dry the DNA pellet under vacuum.

10. Resuspend the pellet in 50 µl of 1× ligase buffer (see Procedure 30).

11. Remove 1–2 µl and save as the preligation control.

12. Add 1 µl of T4 DNA ligase (5–10 units) and gently mix.

13. Incubate at 4°C for 6–18 hr.

DAY 2

1. Remove 1–2 µl, mix with the loading solution, and electrophorese as the postligation control along with the preligation control on a 0.7% agarose gel (see Procedure 43, p. 183).

2. Transform the competent cells with 1–5 µl of ligation mix (see Procedure 37, p. 169).

Notes

If a low yield of transformants is obtained, the ligation mix can be extracted with an equal volume of ether. Remove residual ether under reduced pressure. **Caution:** Ether is extremely flammable!

For a detailed discussion of ligation theory, see Maniatis et al. (1982).

PROCEDURE 32
Transduction with a λ Library

For these procedures, we assume that the λ library has been amplified once as a plate stock and its titer is 10^9–10^{10}/ml.

METHOD A: Transduction

DAY 1

1. Prepare a culture tube with 5 ml of TB containing 0.2% maltose.

2. Inoculate with a single colony of the recipient strain and shake or rotate at 30°C overnight.

DAY 2

1. Centrifuge the overnight culture at 1500g for 10 min. Resuspend the cell pellet in 5 ml of 10 mM $MgSO_4$.

2. Centrifuge the suspension at 1500g for 10 min. Resuspend the cell pellet in 2.5 ml of 10 mM $MgSO_4$.

3. In each of six small test tubes, add aliquots of the λ library and cells as follows:

Test tube	Phage lysate	Cells
1	—	0.1
2	5 μl undiluted	0.1
3	5 μl 10^{-2} dilution	0.1
4	5 μl 10^{-4} dilution	0.1
5	5 μl 10^{-6} dilution	0.1
6	5 μl undiluted	—

4. After 5 min at room temperature, add 2.5 ml of molten (45°C) F top agar containing any supplements needed.

5. Pour on an appropriate minimal agar plate, allow the top agar to solidify, and incubate at 37°C for 24–48 hr. Use 30°C if the λ vector contains the cIts857 mutation or if the strain is otherwise temperature sensitive. In this case, incubate for 36–60 hr.

DAYS 3–6

1. Examine the plates for transductants. Tube 1 reveals the number of revertants. Tube 6 is a sterility control for the lysate.

2. The transductants that arise by a single crossover between the λ phage and the chromosome should be lysogens (see Appendix H, p. 241). The transducing phage can be recovered from the transductants as follows:

   a. Purify eight transductants once on minimal plates with proper supplements.

   b. Grow up 5-ml overnight cultures from single colonies of each transductant in L medium at 30°C.

   c. Save 2 ml of each culture for future use. To the remaining 3 ml, add 0.1 ml of chloroform and incubate at 30°C for 5 min with shaking (or use the rotator).

   d. Remove the dead cells by centrifugation and save the supernatant. If the transductants were lysogens formed by homologous recombination, the supernatant will contain 10^3–10^4 phage/ml (see Appendix H).

   e. Determine the transducing titer and plaque-forming titer of each lysate.

   f. Purify several plaques from each lysate and verify that they transduce the specific marker.

3. Prepare high-titer stocks of the purified transducing phage (see Procedure 1 or 2, pp. 89, 91).

Note

Transduction of a heteroimmune lysogen can provide additional homology for integration (see Appendix J, p. 253).

METHOD B: *Direct Identification of Plaque-forming Transducing Phage*

DAY 1

1. Prepare a culture tube with 5 ml of TB containing 0.2% maltose.

2. Inoculate a single colony of the recipient strain and shake or rotate at 30°C overnight.

DAY 2

1. Centrifuge the overnight culture at 1500g for 10 min. Resuspend the cell pellet in 2.5 ml of 10 mM $MgSO_4$.

2. In a small test tube, infect 0.1 ml of cells with sufficient phage from the library to give 500–1000 plaques per plate. Prepare tubes for 4–8 plates. Include a plate of uninfected cells as a control for reversion.

3. Allow the phage to adsorb for 5 min at room temperature and then add 0.4 ml of TB to each tube, followed by 3 ml of molten (45°C) F top agar.

4. Pour the contents of the tubes on appropriate minimal plates.

5. Incubate the plates at 37°C for 24–48 hr. If the λ vector contains the cIts857 mutation incubate the plates at 30°C and allow an additional day of incubation.

6. The lawn will be very faint but still visible due to the TB added in the overlay. Plaques should be distinct. There will be dense growth in the plaques containing wild-type transductants and complemented lysogens. Often, overgrowth of transductants will obscure individual plaques.

DAYS 3–4

1. Pick the transducing phage plaques to 0.2 ml of λ-dil containing 50 µl of chloroform. Mix well.

2. Purify individual transducing phage by streaking or by plating 10 µl and 100 µl of each plaque suspension on appropriate minimal plates as described in Day 2, steps 2–6.

DAYS 5-6

1. Purify transducing phages and prepare stocks (see Procedure 1 or 2, pp. 89, 91).

2. Verify the transduction of a specific marker.

Note

The λ cloning vector carrying a DNA insert that does not contain the gene of interest is an important control for these procedures.

PROCEDURE 33
Selecting Hybrid Phage According to Density in Cesium Chloride

1. Start with a TB plate containing several thousand plaques from a packaging reaction or transfection. Add 2 ml of λ-dil and scrape off the agar overlay with a pipette or glass rod into a centrifuge tube.

2. Add 0.1 ml of chloroform and vortex. Centrifuge at 4500g for 10 min to remove debris. Save the supernatant.

3. The supernatant should contain about 10^9-10^{10} phage. Add sufficient supernatant to give about 10^8 phage to a CsCl solution prepared by mixing 2.7 g of CsCl with 3.5 ml of 10 mM Tris·HCl, pH 7.4, and 10 mM MgSO$_4$. At this point it is important to add two marker phages, λimm434 and λb538imm434cISam7, to each gradient. Use no more than 10^8 of each marker phage per gradient. Remember that the marker phages must express a different immunity than the vector phage.

4. Add this suspension to a centrifuge tube, fill with paraffin oil, and centrifuge at 30,000 rpm in a Beckman SW50.1 rotor (tube capacity=5.0 ml) or its equivalent for 24 hr at 20°C.

5. The distribution of phage in the gradient is determined by puncturing the bottom of the tube and collecting 7-drop fractions (about 60 fractions) into tubes containing 0.1 ml of λ-dil.

6. Titer each fraction by spotting with a capillary tube on lawns of the following host strains on TB plates: (1) nonimmune host (e.g., LE392); (2) an imm434 lysogen (to locate hybrid phages), and (3) an imm21 or immλ lysogen (homoimmune with the vector, to locate marker imm434 phage). Never use the same capillary tube to transfer aliquots to both strains because there are phage present in these lawns. The titer of appropriate fractions can be assayed more quantitatively on indicator strains, and then phage from a desired region of the gradient are isolated by removing aliquots from appropriate fractions and plating them on LE392. Determine marker peaks and draw a standard curve between markers (plot the fraction number versus the genome size in kilobases).

7. Pick selected plaques to 1 ml of λ-dil containing 50 µl of chloroform. To avoid selecting the marker phages, note the plaque morphology or plating ability: the λimm434 phage forms a turbid plaque and the λb538Sam7imm434cI phage forms a clear plaque. It also carries the Sam7 mutation that is suppressed only by *supF*.

Notes

A variety of well-characterized λ density markers exist; some of these are listed below:

Phage	Percent of wild-type λ (48.5 kb)
λ	100
λimm434	97
λimm21	95
λnin5	94.5
λgal49	90.1
λb2	87.1
λgt·λC	85.1
λWES·λB	82.6
λb538imm434cISam7	80.5
λb221	78

PROCEDURE 34
Preparation of Sau3A Partial Digests of Bacterial DNA

I. TRIAL DIGESTIONS

Optimize conditions for partial digestion by performing serial dilutions of the enzyme according to the following protocol.

DAY 1

1. Mix in a 1.5-ml microfuge tube, in this order:

H_2O	up to 150 µl
10× Sau3A buffer	15 µl (see Notes below)
Bacterial DNA	10 µg

2. Label nine 1.5-ml microfuge tubes. Take 30 µl of the mixture from step 1 and pipette it into tube 1, and put 15 µl of the step-1 mixture into each of tubes 2–8. Put the remainder in tube 9. Put the tubes on ice.

3. Add 4 units of Sau3A to tube 1 and mix well. This gives 2 units of enzyme per microgram of DNA. Mix and transfer 15 µl to tube 2. Mix and transfer 15 µl to tube 3. Continue this twofold serial dilution to tube 8. *Do not* add anything to tube 9.

4. Transfer the tubes to a 37°C water bath for 1 hr.

5. Stop the reaction by placing the tubes on ice and adding EDTA to 20 mM.

6. Load samples 1–9 in order on a 0.7% agarose gel (see Procedure 43, p. 200). Include size markers that span a range of 10–15 kb.

7. Run electrophoresis slowly (1–2 V/cm) overnight.

DAY 2

1. Stain the gel in ethidium bromide (5 µg/ml in water) for 5 min. Rinse with distilled water. Photograph the gel, illuminated by UV light, using several exposure times.

2. Using the markers and the photographs, find which tube gives DNA with the maximum intensity in the 10–12-kb region. When the best tube is chosen, calculate how much enzyme per microgram of DNA gave the results. Proceed to scale-up.

II. SCALE-UP

DAY 1

1. Use half of the amount of enzyme per microgram of DNA that gave the optimal yield of 10–12-kb fragments as defined above.

2. Digest 200 µg of bacterial DNA with Sau3A, using the enzyme concentration determined in step 1 and the exact incubation time and temperature from the original experiment.

3. Run an aliquot of the digested DNA (1 µg) on a 0.7% gel to verify the size distribution of fragments (see Procedure 43, p. 183).

4. Extract the DNA once with chloroform:isoamyl alcohol and ethanol-precipitate the DNA from the aqueous supernatant (see Procedures 40, 41; pp. 177, 180). Resuspend the precipitate in 200 µl of TE (10 mM Tris·HCl, 1 mM EDTA, pH 8.0).

5. Pour a 0.5% agarose gel and load each of the 10 center slots with 20 µl of the DNA. Put markers on both sides at least 5 slots away from the Sau3A-digested DNA. Run electrophoresis slowly 1–2 V/cm overnight.

DAY 2

1. Cut off both sides of the gel containing the markers and stain them with ethidium bromide.

2. Reassemble the gel and examine it under UV light. Using a razor blade, cut out a long strip of unstained gel corresponding to 10–12 kb as determined from the stained fluorescent markers. **Caution:** Wear protective glasses and gloves to prevent UV burns. Stain the remaining gel with ethidium bromide to verify that the proper sample was removed.

3. Place the gel slice into a dialysis bag filled with sterile electrophoresis buffer. Wear gloves.

4. Squeeze out most of the buffer, leaving just enough to cover the gel slice. Tie or clamp the dialysis bag just above the gel slice, eliminating all air bubbles.

5. Place the dialysis bag in a horizontal electrophoresis tank containing a shallow layer of electrophoresis buffer. Orient the bag perpendicularly to the current path to permit efficient electrophoresis of DNA out of the gel.

6. Turn on the current and run electrophoresis for 2–3 hr at about 100 V. The DNA will migrate out of the gel and into the buffer in the bag.

7. Reverse the polarity of the current for 1 min to remove DNA that is sticking to the bag.

8. Open the bag and carefully recover all the buffer in the bag, using a pasteur pipette. Wash the inside of the bag with a small volume of buffer and combine with the initial bag fluid.

9. Stain the gel slice with ethidium bromide and examine under UV light. All the DNA should be eluted.

10. Extract the dialysis bag fluid with phenol/chloroform (see Procedure 40).

11. Precipitate the DNA with ethanol (see Procedure 41).

Alternative Method for Extraction of DNA from Agarose

There is a relatively simple and rapid method for recovery of DNA fragments from agarose gels that is an alternative to elecrophoretic elution. This method is based on the fact that if a piece of agarose gel is frozen, the gel will collapse when centrifuged. Since DNA has little affinity for the agarose, it will remain in the extruded buffer following centrifugation. The method works well for all sizes of DNA fragments, giving excellent recoveries. The phenol extraction/ethanol precipitation step is necessary to eliminate contaminants that would interfere with subsequent manipulation of the DNA.

This protocol was developed by S. Benson and R. Zagursky (National Cancer Institute, Frederick Cancer Research Center).

1. Cut a gel slice containing the desired DNA fragments. Blot excess buffer from the surface of the gel slice and place it in a microfuge tube. Cut the slice into small pieces and distribute to more than one microfuge tube, if necessary. Add 200 μl of phenol. Vortex the sample.

2. Freeze the sample in a dry ice–ethanol bath for 5 min.

3. Centrifuge in a microfuge (~12,000g) for 15 min to pellet the agarose.

4. Remove the top aqueous layer. This is the extruded buffer, which contains the recovered DNA.

5. Add 200 µl of TE (10 mM Tris·HCl, pH 8.0, 1 mM EDTA) to the agarose-phenol remaining in tube.

6. Vortex to reextract and centrifuge for 15 min to separate the layers.

7. Remove the aqueous layer and combine with the sample recovered in step 4.

8. Phenol-extract the aqueous layer twice with an equal volume of phenol (see Procedure 40, p. 177).

9. Precipitate the DNA with ethanol (see Procedure 41, p. 180).

Notes

The recovery of DNA from gels by electroelution is a function of its molecular weight. The larger the DNA fragment, the poorer the recovery. In this procedure, you should expect 30–50% recovery of 10–12-kb fragments.

Prepare dialysis tubing by boiling precut lengths in 2% sodium bicarbonate and 1 mM EDTA for 10 min. Rinse the tubing in distilled water and boil again for 10 min in distilled water. Cool and store submerged in sterile water in the refrigerator. Always keep the tubing free of possible nuclease contamination. Wear gloves when handling tubing.

10× Sau3A buffer

    500 mM NaCl
     60 mM Tris·HCl, pH 7.5
     50 mM MgCl$_2$
      1 mg/ml bovine serum albumin

Sau3A and MboI recognize the same sequence (GATC). Sau3A is insensitive to adenine methylation, whereas MboI will not efficiently cleave methylated DNA.

The procedure outlined here is designed to produce overlapping fragments of bacterial DNA that can be cloned in the λD69 vector (see Appendix I, p. 244). The objective is to obtain 10–12-kb fragments from a partial Sau3A digest of bacterial DNA. For most bacteria, a library of about 2500 transducing phage made from these fragments should represent the entire genome.

PROCEDURE 35
Using Helper Phages to Construct Lysogens of λD69 Hybrids

DAY 1

1. The helper is λb2cIts857 or λb2imm434cIts and should be about 10^{10}/ml. The λD69 hybrid phage should be at a concentration of 10^9/ml or greater.

2. Prepare a culture tube with 5 ml of TB containing 0.2% maltose.

3. Inoculate from a single colony the cells to be lysogenized and shake or rotate at 30°C overnight.

DAY 2

1. In a small test tube, add 50 μl of helper phage and 50 μl of λD69 hybrid phage. Then add 50 μl of overnight cells and mix well.

2. Incubate for 10 min at room temperature. During this period, spread 10^9 selector phage (λint6red3imm21cI) on several L plates. This selector phage will kill nonlysogenic cells but not lysogens carrying a prophage expressing imm21. The red3 and int6 mutations ensure that the selector cannot integrate by phage-mediated recombination.

3. On each prepared L plate, streak a loopful of infected cells from step 2. Incubate at 30°C overnight.

DAY 3

1. Pick single colonies to droplets of λ-dil and streak from these on two L plates. Incubate one at 30°C and one at 42°C. Lysogens that do not grow at 42°C have acquired a helper prophage. Select strains that grow at both temperatures. Verify that these cells are immune to λint6red3imm21cI and sensitive to λb2cIts857 by cross-streaking at 30°C (see Procedure 6, p. 99). Such cells should be lysogens of the λD69 recombinant integrated by site-specific recombination at the attB site between the gal and bio operons.

PROCEDURE 36
Selecting λ Hybrid Phages Constructed with λDamsrIλ3

DAY 1

1. Prepare a culture tube with 5 ml of TB containing 0.2% maltose.

2. Inoculate with a single colony of an appropriate host strain and shake or rotate at 30°C overnight.

DAY 2

1. Centrifuge the overnight culture at 1500g for 10 min. Resuspend the cell pellet in 2.5 ml of 10 mM MgSO$_4$.

2. Mix an aliquot of an in vitro packaging reaction (see Procedure 39, p. 173) that yields about 500 plaques (see Notes below) with 50 µl of cells in MgSO$_4$.

3. Incubate for 10 min at room temperature. Add 3 ml of molten (45°C) TB top agar, mix, and then pour on a TB plate.

4. After 5 min, transfer the plates to 30°C and incubate overnight.

5. Prepare overnight cultures of the following three strains: LE392, N3098 (lig7ts), and 594.

DAY 3

1. Each plate should contain about 500 individual plaques. Purify plaques by removing an agar plug containing the plaque with a pasteur pipette or a capillary pipette (75-mm blue-tip, nonheparinized hematocrit tubes; Fisher Scientific) and by transferring the plug to 1 ml of λ-dil.

2. Add 50 µl of chloroform and vortex the solution. The phage can be stored at 4°C in this state for many months with no loss of titer. Expect about 10^5 phage from a normal-sized plaque.

168 Procedure 36

3. To distinguish parental phages from hybrid phages, spot a 50-μl aliquot of the λ-dil plaque suspension on a series of TB plates containing top agar overlays with the following bacterial strains: LE392, N3098, and 594.

4. After the spots are dry, incubate the plates overnight at 30°C.

DAY 4

1. Each spot on LE392 should give 20–200 plaques.

2. All hybrid phages will not plate or will form no better than pinpoint plaques on the *lig*7ts strain (N3098). A hybrid phage containing an insert >1.9 kb in size does not form plaques on the 594 strain. A hybrid phage with an insert less than 1.9 kb differs in its properties only in that it forms plaques on 594 with the same efficiency as on the LE392 lawn. The parental vector plaques normally on all these strains. It is important to include a spot from a resuspended plaque of the vector on each plate as a control.

Notes

Following *Eco*RI restriction and ligation of a mixture of 0.1 μg of λ*Dam*srIλ3 DNA and 0.2 μg of foreign DNA, in vitro packaging should yield 10^5–10^6 plaques. For DNA transfection of such a mixture, expect 5-fold to 50-fold fewer plaques.

PROCEDURE 37

Transformation of Cells Treated with Calcium Chloride

I. PREPARATION OF CELLS

DAY 1

1. Grow the recipient strain in 500 ml of L medium to $OD_{600}=0.2$.

2. Chill cells thoroughly, for at least 15–20 min on ice. Transfer to cold centrifuge bottles. Keep the cells at 0–4°C for the remaining steps of the procedure.

3. Centrifuge at 2000g for 15 min at 4°C.

4. Resuspend the cell pellet in 200 ml of cold, filter-sterilized 100 mM $CaCl_2$.

5. Incubate the suspension for 20 min on ice.

6. Centrifuge again as in step 3.

7. Resuspend the cell pellet in 5 ml of cold, filter-sterilized 100 mM $CaCl_2$.

8. Store on ice for 20–24 hr.

DAY 2

1. Mix the chilled cell suspension and add sterile glycerol to a final concentration of 10%. Add glycerol slowly with gentle mixing on ice. (When glycerol dissolves in aqueous solutions, heat is released.)

2. Pipette 0.2-ml aliquots into sterile microfuge tubes for storage at −70°C. Keep the tubes on ice. Freeze rapidly in liquid nitrogen and store at −70°C or colder. Each aliquot is sufficient for one transformation.

II. TRANSFORMATION

1. Remove the required number of tubes and thaw cells on ice.

2. Add the DNA solution to the cells and mix gently. Use a minimum volume of DNA solution (<10 µl) for best results.

3. Incubate for 30–60 min on ice.

4. Heat-shock the cells by transferring the tubes to 37°C for 2 min.

5. Add 1.0 ml of L medium and incubate for 30 min at 37°C if phenotypic expression is required. Otherwise, go directly to step 6.

6. Spread the transformation mixture on the selective medium.

7. Incubate plates at 37°C overnight.

Notes

Many factors influence transformation of *E. coli* by plasmid DNAs. The above procedure is very simple and adequate for most needs. A comprehensive study of plasmid transformation has been made by Hanahan (1983).

PROCEDURE 38
Transfection of Cells Treated with Calcium Chloride

I. PREPARATION OF CELLS

DAY 1

1. Prepare a culture tube with 5 ml of TB.

2. Inoculate with a single colony of an appropriate host (e.g., LE392) and rotate or shake at 30°C overnight. These cells will be treated with $CaCl_2$ on Day 2.

3. Prepare a second culture tube with 5 ml of TB containing 0.2% maltose.

4. Inoculate with a single colony of an appropriate host (e.g., LE392) and shake or rotate at 30°C overnight. These cells will be used to titer the phage following transfection.

DAY 2

1. Add 1 ml of the overnight culture grown without maltose to 100 ml of L medium and grow with shaking to $OD_{600}=0.4$.

2. Add 50 ml of cells to each of two centrifuge tubes. Collect the cells by centrifugation at 1500g for 10 min. Decant and discard supernatant.

3. Resuspend each cell pellet in 50 ml of TNa (25 mM Tris·HCl, pH 7.5, and 10 mM NaCl) and centrifuge once again as in step 2. Decant and discard supernatant.

4. Resuspend each cell pellet in 25 ml of ice-cold TNaC (TNa plus 50 mM $CaCl_2$) and keep on ice for 20 min.

5. Centrifuge at 1500g for 10 min and resuspend each cell pellet in 4.15 ml of ice-cold TNaC and keep on ice for 20 min. These cells are now competent for uptake of λ DNA.

II. TRANSFECTION

1. Using glass tubes, add 0.2 ml of cells to 0.1 ml of DNA (~0.1 µg).

2. Add 0.1 ml of TNaC. Mix gently.

3. Incubate on ice for 30 min.

4. Heat-shock for 2 min in a 45°C water bath.

5. Add 1 drop of the overnight culture grown in TB with maltose.

6. Add 3 ml of top agar and pour on a TB plate.

7. Incubate the plate at 37°C overnight.

Notes

Efficiency should be about 10^5-10^6 plaques per microgram of DNA. If glycerol is added to 12%, cells from step 5 can be quick-frozen in liquid nitrogen and stored for future use. To use, thaw cells in cold water and proceed with transfection. This procedure is based on the method of Mandel and Higa (1970).

It is also possible to use $CaCl_2$-treated cells prepared according to Procedure 37 (p. 169) for transfection. In this case use 0.1 ml of thawed, competent cells and add 0.1 ml of TNaC prior to adding DNA. Proceed with transfection as described above.

PROCEDURE 39
In Vitro Packaging of λ DNA

I. PREPARATION OF CELL EXTRACTS

Freeze-Thaw Lysate (FTL)

Prechill all reagents, tubes, and rotors.

DAY 1

1. Streak a colony of BHB2688 on two L plates. Incubate one plate at 30°C and one at 42°C.

DAY 2

1. Examine the plates. There should be no growth on the 42°C plate and good growth on the 30°C plate.

2. Take a heavy inoculum (several colonies) from the 30°C plate to inoculate 500 ml of L medium in a 2-liter flask.

3. Shake at 30°C until the cell density is about 3×10^8/ml (OD_{600}=0.4–0.6). This takes 4–5 hr.

4. Raise the temperature of the culture to 43°C by rapidly swirling the flask in a 90°C water bath; monitor the temperature of the culture with an alcohol-sterilized thermometer. Immediately put the culture into a 43°C bath and shake vigorously for 15 min. Add ice to the water bath to bring the temperature to 38°C and shake *vigorously* for 2.5 hr. Avoid splashing the water-bath liquid into the flask.

5. Chill the flask in ice water. Pour 250 ml of culture into each of two ice-cold centrifuge bottles. Centrifuge for 15 min at 10,000g at 4°C. Pour off the supernatant, removing the last drops with a pasteur pipette and sterile absorbent paper.

6. Keep the centrifuge bottles with cell pellets on ice. Add 0.3 ml of ice-cold 10% sucrose, 50 mM Tris·HCl, pH 7.4, to the first bottle and resuspend the cell pellet uniformly. Do not vortex. Transfer the entire suspension to the second bottle and resuspend the next cell pellet uniformly. Rinse the first bottle with 0.3 ml of ice-cold 10% sucrose, 50 mM Tris·HCl, pH 7.4, and transfer rinse to the second bottle containing both resuspended cell pellets. Mix with a pipette until the suspension is smooth, with no cell clumps.

7. Distribute this semiliquid paste into two *ice-cold* 10-ml ultracentrifuge tubes (0.5 ml/tube). These tubes must be able to withstand freezing in liquid nitrogen.

8. Add 30 µl of fresh, *cold* lysozyme (2 mg/ml, 0.25 M Tris·HCl, pH 8.0). Mix rapidly by hand (do not vortex) and **immediately** freeze in liquid nitrogen. Keep in liquid nitrogen for 15 min. Thaw slowly by immersing the tubes in ice until the paste looks semiliquid and viscous. Then add 125 µl of M1 buffer.

M1 buffer

Make fresh, just before use; add in this order:

H_2O	110 µl
β-mercaptoethanol (14.33 M)	1 µl
0.5 M Tris·HCl, pH 7.4	6 µl
0.05 M spermidine, 0.1 M putrescine (adjust to pH 7 with 1 M Tris Base)	300 µl
1 M $MgCl_2$	9 µl
0.1 M ATP (neutralized with KOH)	75 µl

Mix *gently*, again always *on ice*, and centrifuge at 35,000 rpm in a Beckman 50Ti rotor for 25 min at 2°C. Keep the tubes on ice. Label 75 cryosostorage tubes (e.g., Nunc screw-cap vials) and keep them cold on ice. Add 25 µl of the supernatant to each of the 75 cryostorage tubes. As you add the 25-µl sample, screw the cap tightly and freeze immediately in liquid nitrogen. Work as rapidly as you can.

9. Store the extracts in liquid nitrogen.

Sonic Extract (SE)

Prechill all reagents, tubes, and rotors.

DAY 1

1. Streak a colony of BHB2690 on two L plates. Incubate one plate at 30°C and one at 42°C.

DAY 2

1. Follow steps 1–5 for Day 2 of the freeze-thaw lysate procedure above.

2. Resuspend the cell pellets, as described above in step 6 of Day 2, with a total of 3 ml of *ice-cold* Buffer A (20 mM Tris·HCl, pH 8.0, 3 mM $MgCl_2$, 10 mM β-mercaptoethanol, 1 mM EDTA) and transfer to an *ice-cold*, clear plastic tube or glass Corex tube (capacity, 30 ml).

3. Place the tube in a beaker with ice and salt water. Sonicate with the microtip at full power with several 3-sec bursts with cooling pauses of 15 sec in between (never sonicate unless the tip is fully immersed); this is essential to *avoid foaming*. The suspension should become very viscous, then lose some of this viscosity. It will become iridescent. Anywhere from 8 to 20 bursts may be needed. *Keep the tube cold*. The final product will be the consistency of cream.

4. Centrifuge the tube for 10 min at 6000 rpm in an ice-chilled Sorval SS-34 rotor or its equivalent at 2°C. If sonication is done properly, the pellet should be small. Add 0.6 ml of M1 buffer to the supernatant and distribute the entire supernatant into *ice-cold* cryostorage tubes at 25 μl per tube. Cap the tubes and drop them **immediately** into liquid nitrogen.

5. Store the extracts in liquid nitrogen.

II. IN VITRO PACKAGING REACTION

1. Thaw FTL and SE by placing on ice for 15–30 min. The extracts are very viscous.

2. Gently mix together, avoiding foaming, in a microfuge tube at room temperature *in this order:*

SE	5 μl
DNA	up to 500 ng (see Notes below)
FTL	25 μl

3. Incubate for 1 hr at room temperature.

4. Add 300 μl of λ-dil and 50 μl of chloroform. Vortex briefly and centrifuge in a microfuge for 3 min to separate the phases. The chloroform is at the bottom. Remove the aqueous, top layer to a clean tube.

5. Titer as a phage lysate on cells grown in 0.2% maltose (see Appendix G, p. 239). Store primary packaging lysate at 4°C in the dark.

Procedure 39

Notes

Use no more than 3 µl of DNA in the in vitro packaging reaction.

Efficiencies have ranged from 10^6-10^8 plaques per microgram of uncut λ DNA. After digestion and religation, this may fall by a factor of 10–100.

Both SE and FTL can be refrozen, but with loss of activity (up to 50%).

Do not package the DNA directly from ligation mixtures. Extract the ligation mixture with an equal volume of chloroform/isoamyl alcohol, followed by ethanol precipitation (see Procedures 40, 41; pp. 177, 180). Resuspend the DNA precipitate in an appropriate volume of TE (10 mM Tris·HCl, 1 mM EDTA, pH 8.0).

Packaging extracts contain the concentrated remains of lysed bacteria as well as other products of phage development. Most notable are large numbers of free phage tails. Consequently, the amount of a packaging reaction that can be plated directly is limited. In fact, do not plate more than 20 µl of packaging reaction or you will inhibit plaque formation. Be sure to use a recipient grown in TB containing 0.2% maltose as the plating bacteria. This will ensure the maximum number of λ receptors on the cell surface.

Phages in packaging reactions can be purified quickly by glycerol step-gradient centrifugation (see Procedure 4, p. 95).

Packaging extracts are most stable at −70°C or in liquid nitrogen. They lose activity at −20°C.

Both strains, BHB2688 and BHB2690, form slightly mucoid colonies.

Several types of in vitro packaging extracts are available from various vendors. These may suffice for certain purposes.

This procedure is a modification of published protocols (Enquist and Sternberg 1979; Hohn 1979). It was developed by N. Sternberg (National Cancer Institute, Frederick Cancer Research Facility).

PROCEDURE 40

Phenol/Chloroform Extraction of DNA Samples

I. PREPARATION OF EXTRACTION SOLUTIONS

Phenol

1. Carefully distill phenol at 160°C. Store purified phenol in 100-ml aliquots at −20°C until needed.

2. As needed, remove phenol from the freezer and warm to room temperature. Melt the phenol crystals in a 70°C water bath.

3. Shake the melted phenol with an equal volume of distilled water and allow to stand at room temperature until the phases separate.

4. Remove the lower, water-saturated phenol layer to a fresh, sealable glass or polypropylene container. Phenol at this stage can be stored for several months at 4°C.

5. On the day of use, remove only the required amount of phenol and mix with an equal volume of 1 M Tris·HCl, pH 8.0.

6. Centrifuge at 1000g for 5 min to separate the layers. Discard the top, aqueous layer.

7. Save the lower, phenol layer and mix with an equal volume of 0.1 M Tris·HCl, pH 8.0.

8. Centrifuge at 1000g for 5 min to separate the layers.

9. Remove most of the top, aqueous layer and save the phenol layer. Store on ice or at 4°C for use that day.

   Caution: Phenol is dangerous and can cause severe burns. Always wear gloves and safety glasses. If phenol touches your skin, rinse quickly with large quantities of water, followed by a wash of soap and water. Do *not* rinse with ethanol. If large areas of skin surface are involved, seek immediate medical attention.

Chloroform: Isoamyl Alcohol (24:1 v/v)

1. In a large screw-capped bottle, add 240 ml of chloroform and 10 ml of isoamyl alcohol.

2. Mix and wrap the bottle in aluminum foil to protect it from light.

3. Store, tightly capped, at room temperature.

II. DNA EXTRACTION

1. Mix the DNA solution with an equal volume of phenol from part I, step 9 in a 1.5-ml microfuge tube.

2. Snap the cap shut and invert the tube rapidly until an emulsion forms.

3. Centrifuge for 1–5 min in a microfuge (~12,000g) at room temperature to separate the layers.

4. Use a pasteur pipette or a micropipette tip to transfer the upper aqueous phase to a clean, 1.5-ml microfuge tube. Discard the interface and the lower phenol layer. (For optimum recovery, the interface and phenol layer can be reextracted with an equal volume of buffer and centrifuged as in step 3 above, and the aqueous layer can then be mixed with that obtained after 3.)

5. Add an equal volume of (phenol):(chloroform:isoamyl alcohol) (mixed 1:1). Repeat steps 2–4.

6. Add an equal volume of chloroform:isoamyl alcohol and repeat steps 2–4.

7. Recover the DNA by ethanol precipitation as described in Procedure 41 (p. 180).

Notes

Although this procedure is designed for 1.5-ml microfuge tubes, larger polypropylene tubes can be used (e.g., the 12×75 mm, snap-cap, Falcon 2063 tube; this tube can be centrifuged at 3000 rpm for 5 min in standard bench-top centrifuges to separate the phenol and aqueous layers).

Always wear gloves during the phenol/chloroform extraction procedure.

Water-saturated ethyl-ether can be used to remove traces of phenol or chloroform from the final product. However, in the experiments described in this manual, ether extraction is not necessary. Because ether is highly volatile and flammable, extreme **caution** must be used.

Phenol solutions in the commonly used Tris buffers tend to turn yellow quickly and should then be discarded. It is prudent to prepare buffered phenol fresh for each use. Phenol in water does not show this effect and can be stored conveniently in this form if protected from light and kept at 4°C. The antioxidant 8-hydroxyquinoline can be added to phenol at a concentration of 0.1% to prevent formation of undesirable compounds. It imparts a yellow-orange color, which is useful in identifying the phenol layer.

PROCEDURE 41
Ethanol Precipitation of DNA

1. Transfer a known volume of DNA in TE (10 mM Tris·HCl, 1 mM EDTA, pH 8.0) to a 1.5-ml microfuge tube. Use no more than 0.4 ml.

2. Add 0.1 volume of 3 M sodium acetate and mix well.

3. Add exactly 2 volumes of ice-cold 95% ethanol and mix well.

4. Place the tube in a dry ice–ethanol bath for at least 5 min. Beware—some ink used for labeling tubes is soluble in ethanol. A convenient tube holder is a flat piece of Styrofoam with holes the size of microfuge tubes punched in it. This floats directly in a beaker of dry ice and ethanol.

5. Collect the DNA precipitate by centrifugation for 5 min in a microfuge ($\sim$12,000g).

6. Pour off the supernatant, invert the tube on a layer of absorbent paper, and tap gently to drain the residual ethanol.

7. Final traces of ethanol can be removed by a 1–2-min treatment in a vacuum dessicator or by a brief exposure to a stream of nitrogen gas in a hood.

8. Resuspend the pellet (often invisible) in the desired buffer.

Notes

This procedure is designed for use with 1.5-ml microfuge tubes with snap caps. As always, wear gloves to protect DNA from contamination when using this procedure.

It is sometimes convenient to use carrier tRNA to coprecipitate small concentrations of DNA (<0.1 μg/ml). In step 2, add 1 μl of 10 mg/ml yeast tRNA that has been dissolved in water and boiled for 5 min.

In step 2, remember that a moderate amount of monovalent cations must be present. The exact concentration is not important so long as it is in the range of 0.1–0.4 M. If the monovalent cation concentration is above 0.4 M, use TE to dilute appropriately.

DNA pellets from step 8 dissolve easily in low-salt buffers like TE. However, warming the tube at 37°C or mixing with a sealed pipette may aid in dissolving the DNA.

For very low DNA concentrations (<0.1 µg/ml in the absence of carrier tRNA), the time in the dry ice–ethanol bath can be extended to several hours. Alternatively, a −70°C freezer can be used.

When very small DNA fragments are being precipitated, the microfuge may not be sufficient. An alternative method uses a Beckman SW41 or equivalent rotor in the ultracentrifuge at 35,000 rpm for 20–30 min (in step 5).

If additional purification seems necessary (i.e., if the DNA cannot be digested with a restriction enzyme or ligation is poor), use the drop-dialysis method (Procedure 42, p. 182). This method removes salt and detergents that often cause these problems.

PROCEDURE 42
Drop Dialysis of DNA Preparations

1. Pour 5–10 ml of dialysis buffer—usually TE (10 mM Tris·HCl, 1 mM EDTA, pH 8.0)—in a sterile, plastic petri dish.

2. Float a 25-mm-diameter, type-VS Millipore membrane (cat. no. VSWP 02500, MF type, VS filter, mean pore size=0.025 μm) *shiny side up* on the dialysis buffer. Allow floating filter to wet completely before proceeding.

3. Pipette the DNA droplet carefully (20–100 μl) onto the center of the filter.

4. Cover the petri dish. Dialysis can be done at room temperature or in the cold. Be careful not to jar the dish or you may experience the thrill of seeing the filter and your sample flip over. Dialyze for at least 1 hr and no more than 4 hr.

5. Carefully retrieve the DNA droplet with a micropipette. Steps 1–5 can be repeated with fresh buffer if more dialysis is required.

Notes

The method is useful for removal of substances that inhibit restriction endonucleases or DNA ligases. It is particularly effective for elimination of SDS or a high salt concentration from DNA preparations.

Step 5 can be tricky for those with shaky hands or poor hand-eye coordination. The filter has a tendency to move briskly around the surface as you touch it with the pipette tip. Practice with buffer droplets to master the technique before you try using a valuable sample.

PROCEDURE 43
Restriction Endonuclease Digestion and Gel Electrophoresis of DNA

Bacterial DNA Digestion[1]

1. Add to 1.5-ml microfuge tubes, in this order:

Distilled H_2O	up to 25.0 μl
10× buffer	2.5 μl
DNA (up to 1 μg)	2.0 μl
Enzyme (10 units/μg DNA)	X μl
Total volume	25.0 μl

   The DNA may be dialyzed prior to digestion (see Procedure 42).

2. Mix gently. Close tubes tightly and incubate at 37°C for 2 hr.

Agarose Gel Electrophoresis

1. Make enough 1× electrophoresis buffer (EB) for the particular electrophoresis apparatus (stock is 20×; see below).

2. Make sufficient 0.7% agarose in 1× EB. Dissolve in a boiling water bath or a microwave oven. When cooled to about 50°C, cast the gel in an appropriate apparatus.

3. Mix in a microfuge tube:

   2.5 μl of bacterial DNA digestion
   5.0 μl of loading solution (see below)
   10–15 μl of 1× EB to make final volume of 20–25 μl

---

[1]This protocol is designed for Experiments 4 and 5 (pp. 33, 39). When digesting plasmid or phage DNA, less enzyme is required.

Generally, 0.1 µg of DNA is sufficient, but this will vary depending on the number and size of the fragments.

4. Heat in a 70°C water bath for 10 min and then quickly cool on ice.

5. Load the samples in the gel slots, using a micropipette. Include another sample containing DNA restriction fragments of known size as a control.

6. Attach the leads to the power supply. DNA will migrate to the positive pole. Run electrophoresis as fast as conditions will permit. Be careful not to overheat the gel; it can melt.

7. Stain by immersing the gel in a water solution of ethidium bromide (5 µg/ml). Stain for 5–10 minutes and rinse with distilled water.

8. Photograph the gel, using UV light.

Acrylamide Gel Electrophoresis of DNA Fragments

1. To make 50 ml of 6% acrylamide gel solution, combine the following:

    5.0 ml 10× TBE
    7.5 ml acrylamide stock (see below)
    37.5 ml H_2O
    0.5 ml 10% ammonium persulfate (freshly prepared)
    50.0 µl TEMED (N,N,N',N'-tetramethyl-ethylenediamine)

2. Cast the gel in an appropriate apparatus. This formulation should polymerize in 5–10 min.

3. Attach the leads to the power supply. DNA will migrate toward the positive pole. Run the electrophoresis, using 1× TBE, as fast as conditions will permit.

4. Stain and photograph as described above.

Reagents for Restriction Endonuclease Digestions and Electrophoresis

All restriction enzyme buffers should be made to the specifications of the vendor. The buffers are usually made and stored in 10× concentrations at 4°C.

Loading solution

   20% Ficoll
   0.05% bromophenol blue

20× Electrophoresis buffer (EB)

> 800 mM Tris
> 40 mM disodium EDTA
> 400 mM sodium acetate

> 96.9 g Tris base, 15.38 g disodium EDTA, 53.43 g sodium acetate. Stir to dissolve. Adjust the final pH to 7.9 with acetic acid. Add water to 1 liter and autoclave. Store at room temperature.

Acrylamide stock

> Combine 40 g of acrylamide and 1.3 g bisacrylamide in water to give a final volume of 100 ml. Store at 4°C.

> **Caution:** Acrylamide is a neurotoxin. Wear gloves and do not mouth-pipette. Weigh powder in a hood and wear a mask.

10× TBE

> 500 mM Tris
> 500 mM boric acid
> 25 mM disodium EDTA

> 60.55 g Tris base, 30.90 g boric acid, 9.3 g disodium EDTA. Stir to dissolve. Add water to 1 liter. Store at room temperature.

Ethidium bromide solution

> Ethidium bromide solution (5 µg/ml) is made by diluting a 5 mg/ml stock solution in water. The stock solution is stored at 4°C and protected from the light.

> **Caution:** Ethidium bromide is a suspected carcinogen. Always wear gloves when using it and when handling gels that have been stained with it.

Notes

Keep all DNA samples and reaction buffers on ice throughout the procedure. Wear gloves at all times. Use sterile, nuclease-free plastic tubes and pipettes.

PROCEDURE 44
Southern Blot Transfer of DNA

DAY 1

1. Wear gloves throughout this procedure. Run the electrophoresis of the DNA in an agarose gel under desired conditions. Stain the gel in ethidium bromide (5 µg/ml) and photograph using UV light (see Procedure 43, p. 183).

2. Trim and mark the gel for orientation by cutting off a corner with a razor blade.

3. Always using a glass plate as a support, transfer the gel sequentially into the following solutions contained in glass baking dishes:

   a. 500 ml of depurination buffer for 15 min. Rinse briefly in water, then repeat.

   b. 500 ml of denaturation buffer for 15 min. Rinse briefly in water, then repeat.

   c. 500 ml of neutralization buffer for 30 min. Rinse briefly in water, then repeat.

4. Prepare the blotting set up as follows: Cut 12 pieces of Whatman 3MM paper or its equivalent, 1 piece of nitrocellulose membrane (Schleicher & Schuell, BA85, or equivalent) and an 8–10-cm stack of absorbent paper towels to the exact size of the gel. Soak the nitrocellulose membrane first in distilled water at room temperature for 20 min. If the membrane retains any dry areas, bring the water to boiling and then slowly cool down. Soak the membrane next in 20× SSC for 20 min.

5. In a glass baking dish or plexiglass box, place 6 sheets of Whatman 3MM paper from step 4. Pour a small amount of 20× SSC into the dish, just enough to saturate and surround the paper. *Do not submerge the papers*. Smooth out all the air bubbles with a gloved hand. Place the gel on top of the stack of 3MM paper. Without stretching the gel, smooth out all air spaces between paper and gel with a gloved hand. Put 1 drop of ethidium bromide (1 µg/ml) into each well of the gel.

6. Carefully position the precut nitrocellulose membrane (step 4) on top of the gel. DNA transfer begins immediately, so do not move the membrane once it has been placed on the gel. If the placement is unsatisfactory, remove and discard the membrane and try again with a new membrane prepared in step 4. With a gloved hand, smooth out all air spaces between gel and membrane.

7. Place the remaining 6 sheets of Whatman 3MM paper on top of the membrane. Wet gently, but thoroughly, with 20× SSC. Use just enough 20× SSC to saturate the paper. With a gloved hand, smooth out all the air spaces between the paper sheets.

8. Position the precut stack of absorbent paper toweling on top of the 3MM paper. Place a glass plate large enough to cover the toweling on top of the stack. Place a 1-kg weight on the glass plate. A suitable weight is a flask containing 1 liter of water (see Fig. 20).

9. "Blot" at room temperature overnight.

DAY 2

1. Remove the weight, towels, and 3MM paper above the gel. Carefully remove the nitrocellulose membrane. Place the nitrocellulose membrane between two sheets of 3MM paper and bake at 80°C for 2 hr in a vacuum oven. To assess the efficiency of DNA transfer, the dried gel can be stained with ethidium bromide (5 μg/ml) and compare with the picture of the preblotted gel.

2. Allow the membrane to cool slowly to room temperature. Prehybridize and hybridize using desired conditions (see Procedure 46, p. 191) or store the membrane between 2 sheets of 3MM paper in a cool, dry place.

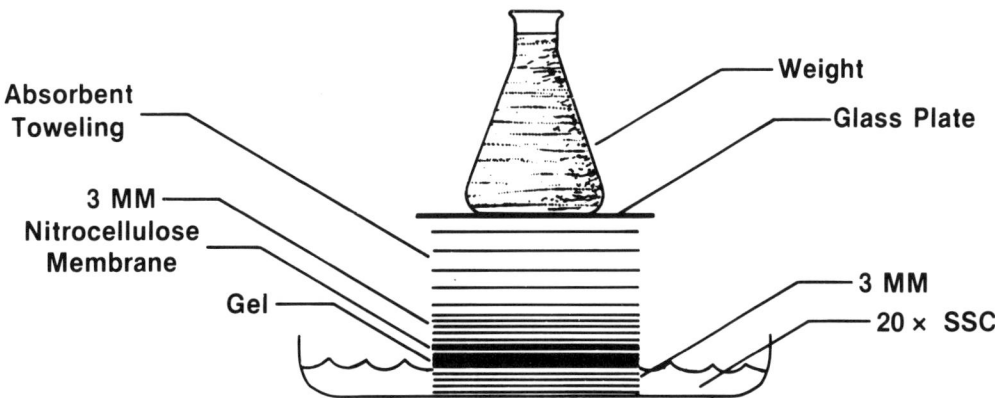

Figure 20

Southern Blot Reagents

Depurination buffer

    0.25 M HCl

    21 ml conc. HCl in 1 liter H_2O

Denaturation buffer

    0.5 M NaOH
    1 M NaCl

    20 g NaOH, 58.44 g NaCl, H_2O to 1 liter.

Neutralization buffer

    1.0 M Tris·HCl, pH 8.0
    0.6 M NaCl

    121 g Tris base, 34.8 g NaCl, 55 ml conc. HCl, H_2O to 1 liter. Adjust pH to 8.0 with HCl, if necessary.

20× SSC

    3.0 M NaCl
    0.3 M sodium citrate

    175.3 g NaCl, 88.2 g sodium citrate, dissolve in H_2O up to 1 liter. Adjust to pH 7.0 with concentrated NaOH.

Notes

Never handle nitrocellulose membranes with your bare hands. Always wear gloves. Oil from your hands will ruin the membranes.

The transfer of DNA from agarose gels to membranes (either nitrocellulose or chemically activated paper) for hybridization studies is a powerful technique that has had a great impact on many areas of research. The procedure described here is based on the original method of Southern (1975)—hence the eponyms Southern blot or Southern transfer. Many modifications of this protocol have been described, including electrophoretic transfer of DNA from gels.

The amount of buffer in the bottom of the dish (Day 1, step 5) should be just enough to surround the initial 6 sheets of 3MM paper. It should not rise to the level of the gel.

It is important that the absorbent paper toweling does not dip down into the reservoir of buffer. This will draw liquid around the gel rather than through the gel, as intended.

PROCEDURE 45
Dehydration of Agarose Gels for DNA-DNA Hybridization

1. Run the electrophoresis of the DNA samples in an agarose gel under the desired conditions (see Procedure 43, p. 183). The best results have been obtained with gels no larger than 15 cm × 20 cm and no more than 5 mm thick. Gels of 0.7%–1% agarose are most commonly used.

2. Immerse the gel in ethidium bromide (5 µg/ml) for 5 min to visualize the DNA. Rinse the gel in distilled water and then photograph it, using UV light.

3. Denature the DNA by soaking the gel for 1 hr at room temperature in 50 mM NaOH. Use a glass baking dish and enough NaOH to cover the gel.

4. Immerse the gel for 30 min at room temperature in distilled water.

5. Carefully transfer the wet gel to Whatman 3MM paper, or its equivalent, cut to be 2 cm longer and wider than the gel.

6. Place the gel and paper in a gel drier as follows. First, cover the bottom of the gel drier with a sheet of 3MM paper cut to the exact size of the drier. Then place the gel and 3MM paper from step 5 in the middle of the gel drier (gel is up). Cover the gel with plastic wrap. Be sure that the plastic wrap covers the entire gel drier. Put the rubber flap of the drier directly over the plastic wrap.

7. Dehydrate the gel thoroughly. For a 5-mm-thick, 0.7–1.0% agarose gel, dry for 30 min with no heat, then for 30 min with heat (60°C), and finally for 15–20 min with no heat.

8. Interrupt the vacuum and turn off the gel drier. Remove the 3MM paper onto which the agarose gel is now dried to a thin film. A properly dried gel will appear to be embedded in the 3MM paper. Be careful when peeling off the plastic wrap. Start from one corner and pull back slowly. Do not tear the gel. Gently place the 3MM paper with the dried-on gel into a dish of distilled water. The 3MM paper will slowly sink to the bottom of the dish as the dried agarose gel slightly rehydrates and floats to the surface; this takes about 15 min. The gel will be about the thickness of a dialysis membrane.

9. Prehybridize, hybridize, wash, and autoradiograph the gel under the desired conditions (see Procedure 46).

Notes

This procedure is designed for use with commercial gel driers. A large plastic sheet that fits in the drier is not used.

The procedure is a modification by investigators at Molecular Genetics, Inc., of a protocol from S. Tsao, C. Brunk, and R. Pearlman (Department of Biology, York University, Ontario, Canada). A similar protocol is described by Shinnick et al. (1975).

The most common factor contributing to a high background upon hybridization of the dried gel to radioactive DNA probes is the failure to dry the gel thoroughly. Longer drying times may be required for thicker gels or weaker gel drier vacuum systems.

PROCEDURE 46
DNA-DNA Hybridization

A previously prepared Southern blot (see Procedure 44, p. 186), dehydrated agarose gel (see Procedure 45, p. 189) or plaque blot (see Procedure 47, p. 196) is incubated first in a prehybridization solution for at least 3 hr (up to 16 hr is acceptable and recommended for the dehydrated gel). The hybridization reaction is then incubated for 4–16 hr in a similar solution containing a radiolabeled DNA "probe." For both incubations, stringent conditions are achieved in 50% formamide at 42°C. The hybridization reaction is followed by washes of decreasing salt concentration and increasing temperature (up to 65°C in some instances). These conditions allow detection of strong homology between the probe and the DNA in question, minimizing background hybridization of the probe to DNA sequences of weak homology.

I. PREHYBRIDIZATION

Prehybridization solution

5× SSC	2.5 ml	20× SSC (see Procedure 44)
50% formamide	5.0 ml	100% formamide
5× Denhardt's solution	0.5 ml	100× Denhardt's solution (see below)
50 mM sodium phosphate buffer, pH 6.5	1.0 ml	0.5 M sodium phosphate buffer, pH 6.5
1 mg/ml sonicated salmon sperm DNA (sssDNA)	1.0 ml	10 mg/ml sssDNA
Total	10.0 ml	

1. Mix together the above reagents in the order given.

2. Using 1 ml of this mixture, wet the inside of a heat-resistant, sealable cooking pouch that has been opened on three sides.

3. Place the nitrocelluose membrane or the dehydrated gel into the bag.

4. Seal two edges by heating with a bag sealer.

5. Add the remaining 9 ml of solution.

6. Squeeze out all bubbles and seal the fourth side.

7. Encase this in a second bag and seal well. Immerse in a 42°C water bath.

II. HYBRIDIZATION

1. Prepare 0.5 ml of 10 mg/ml sssDNA and an aliquot of probe DNA (to be determined, based on the specific activity of the probe). Mix them together in a screw-capped polypropylene tube (50 ml capacity) and put into a boiling water bath for 5 min. Do not screw cap on tightly. Cool quickly and thoroughly on ice. This heat denatures the probe DNA.

2. Add the remaining reagents to the denatured probe in the following order:

   2.5 ml 20× SSC
   5.0 ml 100% formamide
   0.1 ml 100× Denhardt's solution (final concentration, 1×)
   0.4 ml 0.5 M sodium phosphate buffer, pH 6.5 (final concentration, 20 mM)
   2.0 ml 50% sodium dextran sulfate (final concentration, 10%)

3. Cut a corner from the bag containing the prehybridization solution.

4. Place the bag on a flat surface and gently squeeze out the liquid by rolling a pipette over its surface.

5. Carefully pipette the hybridization solution in the same bag.

6. Squeeze out all bubbles and seal the corner of the bag.

7. Place into a second bag and seal well.

8. Immerse in a 42°C water bath.

III. POSTHYBRIDIZATION WASH

1. Open the bag at the corner and carefully squeeze the radioactive hybridization mixture into a tube.

2. Cut open the bag on all sides and transfer the wet, radioactive membrane or gel to the following sequential washes in glass baking dishes:

   a. 2× SSC, 1× Denhardt's solution for 15 min at room temperature, then repeat.

   b. 2× SSC, 0.1% SDS for 30 min at room temperature, then repeat three times.

   c. 0.1× SSC, 0.1% SDS for 30 min (or longer) at 50–65°C, then repeat three times. **Caution:** Do not wash a dehydrated gel above 50°C.

   Follow the elution of background radioactivity with a handheld monitor. Wash times may be increased if the background is unacceptable.

3. After the final wash, blot excess liquid from the membrane or gel with absorbent paper. Wrap the membrane or gel in plastic wrap. *Do not forget this step.* The film will stick to the membrane or gel during autoradiography, ruining them for future use unless they are wrapped in plastic.

4. Insert into a casette with X-ray film as follows:

   a. Tape the plastic-wrapped membrane or gel to a backing of Whatman 3MM paper cut to fit in the film holder. You may mark the orientation with radioactive ink (a small amount of ^{32}P mixed with waterproof ink).

   b. Wrap the sample and 3MM backing sheet in plastic wrap.

   c. In the darkroom, place the plastic-wrapped sheet in an X-ray film holder and cover it with a sheet of X-ray film. Kodak X-Omat AR is most versatile because it is a fast, two-sided emulsion film.

   d. Next, place an intensifying screen (Dupont Cronex Lightning-Plus) on the film, with the shiny side down.

   e. Close the film holder, wrap in aluminum foil, and place at −70°C for a time ranging from hours to days, depending on the specific activity of the probe.

   f. Develop the film.

Notes

The association rate of nucleic acids is accelerated in the presence of dextran sulfate because the nucleic acids are excluded from the volume of the solution occupied by the polymer. Therefore, their effective concentrations are increased. Problems may arise because of the viscosity of dextran sulfate and because it can cause high backgrounds.

The use and concentration of dextran sulfate (see Hybridization, step 3) is dependent upon experimental conditions and may affect the signal-to-noise ratios. It is advisable to decrease the concentration of dextran sulfate to 1–2% for hybridization of dried-down gels.

When hybridizing a dehydrated gel, do not use temperatures above 42°C. It is not necessary to agitate the bag unless there are several filters in it. In this case agitation prevents the filters from sticking together.

A weight should be placed on the film holders, if possible, to press the sample firmly to the film, thus ensuring sharp images.

Remove the holder from the freezer and allow it to warm to room temperature before removing the film for developing. If condensation forms on the screens and the sample, wipe off before applying new film.

Because these experiments involve the use of radiolabeled DNA, it is especially important that workers follow recommended laboratory procedures. Wear protective clothing and gloves throughout the entire experiment and dispose of radioactive waste as required.

Reagents for Hybridization

Sodium dextran sulfate (M_r 500,000)

50% (w/v) in sterile water. Heat to 60°C to dissolve. Store at room temperature.

100× Denhardt's solution

Mix together:

2 g Ficoll (M_r 400,000)
2 g PVP (polyvinyl pyrrolidone; M_r 360,000)
2 g BSA (bovine serum albumin, nuclease-free)

Adjust to 100 ml with 3× SSC. Stir to dissolve (overnight if necessary). Aliquot and store frozen at −20°C.

Sonicated salmon sperm DNA (sssDNA)

Use type-III sssDNA, as sodium salt from salmon testes (Sigma).

1. Dissolve 1 g of DNA in 100 ml of 20 mM Tris·HCl, pH 7.5. Let stand overnight at room temperature.

2. Sonicate 25-ml aliquots on ice until the solution has the consistency of milk.

3. Denature the sheared DNA in a boiling water bath for 15 min.

4. Quick-cool on ice. Aliquot and store at −20°C.

5. Determine the fragment size on a 1% agarose gel (see Procedure 43, p. 183). An average fragment size of ~700 bp is ideal.

Sodium phosphate buffer (0.5 M, pH 6.5)

Prepare the following stock solutions:

    A. $NaH_2PO_4 \cdot H_2O$ 69 g/liter
    B. $Na_2HPO_4 \cdot 7H_2O$ 134 g/liter

Mix:

    68.5 ml solution A
    31.5 ml solution B

PROCEDURE 47
Plaque Hybridization

DAY 1

1. Prepare a culture tube with 5 ml of TB containing 0.2% maltose.

2. Inoculate with a single colony of the appropriate strain (see Notes below) and shake or rotate at 37°C overnight.

DAY 2

1. Centrifuge the overnight culture at 1500g for 10 min. Resuspend the cell pellet in 2.5 ml of 10 mM $MgSO_4$. Store cells at 4°C.

2. Prepare plaques for hybridization as follows:

   a. Use dry L plates (at least 2 days old).

   b. Use 0.7% water agar (see Appendix A, p. 217). Melt and cool to 45°C prior to use.

   c. Mix 50 µl of cells from step 1 and an aliquot of phage, diluted to give about 10^4 plaques per plate, in a small test tube.

   d. Incubate at room temperature for 5 min.

   e. Add 4 ml of molten (45°C) water agar and pour on a prelabeled L plate.

3. Incubate the plates at 37°C overnight.

DAY 3

1. Cool the plates containing plaques for 1 hr at 4°C to harden the top agar layer.

2. Place a dry, round nitrocellulose filter on the lawn of cells. (If the filter has a grid pattern, place the printed side down.) Be careful not to let air bubbles form between agar and filter. *Do not move, lift, or reposition the filter*.

3. Leave the filter on the plate for 15 min at 4°C. During the adsorption, mark the filter orientation. To mark the filter and orient it relative to the plate, stab through the filter and into the agar at the edge of each filter with a 25-gauge needle containing waterproof black drawing ink (this can be mixed with 1 µg/ml denatured, unlabeled probe DNA). Apply about 10 µl at each spot. Mark each plate uniquely and nonsymmetrically.

4. Carefully remove the filter from the plate and turn the plaque side up. Be careful not to lift the top agar from plate. Avoid using sharp-tipped forceps; instead, use round-tipped forceps to *gently* lift the filter. Do not break the agar surface. Do not allow the filter to *retouch* the agar surface. Lift off in a smooth, gentle motion. If the agar begins to come up, try to lift the opposite corner and pull the filter up carefully. Wrap the agar plates in parafilm or plastic wrap and store at room temperature for use in step 1 of Day 4.

5. On a large sheet of parafilm or in three glass trays, arrange stacks of three sheets of Whatman 3MM paper. If you assemble the stacks on a large sheet of parafilm, make sure the stacks of 3MM paper are at least 6 in. apart.

6. Saturate the first stack with 0.5 M NaOH containing 1.5 M NaCl. Do not allow buffer to pool on the 3MM paper. Lay the filter (plaque side up) on the stack for 10 min.

7. Saturate the second stack with 1 M Tris·HCl, pH 7.5, containing 1.5 M NaCl and transfer the filter to this stack for 10 min.

8. Saturate the third stack with 2× SSC and transfer the filter to this stack for 5 min.

9. Blot the filter with clean 3MM paper to dry.

10. Bake at 80°C for 2 hr.

11. Prehybridize and hybridize as usual (see Procedure 46, p. 191) and put on film.

DAY 4

1. Wash the hybridized filters and place them on X-ray film (see Procedure 46).

DAY 5

1. Develop the X-ray film after 1–2 days, depending on the specific activity of the DNA probe.

2. Align the developed X-ray film with the original plate and locate the "positive" plaques.

3. Pick plaques showing positive hybridization into a fresh, top agar lawn on a TB plate seeded with the host strain (saved from Day 2).

4. Incubate this plate at 37°C overnight.

DAYS 6–7

1. Make a new filter from this plate, rehybridize, and confirm a positive stab.

2. Pull an agar plug from the stab area, using a pasteur pipette, and put it in 1 ml of λ-dil. Store at 4°C.

3. Purify the phage by streaking or replating.

4. Make phage stocks (see Procedure 2, p. 91).

PROCEDURE 48
Nick Translation of DNA

1. Prepare DNase I as follows:

Stock	Volume	Final concentration
H$_2$O	10 ml	—
1 M Tris·HCl (pH 7.5)	100 µl	10 mM
1 M MgCl$_2$	50 µl	5 mM
DNase I (1 mg/ml)	1 µl	100 ng/ml

Use RNase-free DNase I (cat. no. DPFF, 2000–2600 units/mg; Worthington). Purchase 5-mg lots and resuspend the entire lot in 50% glycerol, 10 mM Tris·HCl, pH 7.5. Store at −20°C in 100-µl aliquots.

2. Incubate for 1 hr on ice.

3. Prepare the following reaction mix (final volume, 50 µl).

Stock	Volume	Final concentration
1 M Tris·HCl (pH 7.5)	2.5 µl	50 mM
0.1 M MgCl$_2$	2.5 µl	5 mM
β-Mercaptoethanol (100 mM = 7.1 µl/ml)	5.0 µl	10 mM
Bovine serum albumin (1 mg/ml)	5.0 µl	100 µg/ml
Probe DNA in H$_2$O	15.0 µl	∼ 1–2 µg DNA
[α-^{32}P]dTTP	5.0 µl	50 µCi (10 mCi/ml, ∼ 6000 Ci/mmole)
[α-^{32}P]dCTP	5.0 µl	50 µCi (10 mCi/ml, ∼ 6000 Ci/mmole)
dATP (100 µM)	5.0 µl	10 µM
dGTP (100 µM)	5.0 µl	10 µM

4. Add 1 μl of diluted DNase I for DNA fragments greater than 5 kb and 2 μl for smaller DNA fragments. Use 0.25–1 μg of DNA per reaction.

5. Incubate for 10 min at 14°C.

6. Add 0.5 μl of DNA polymerase I (2.5 units).

7. Incubate for 1 hr at 14°C.

8. Remove unincorporated, labeled nucleoside triphosphates, using minicolumns (see Procedure 50, p. 203).

Notes

This protocol uses limited DNase I digestion of DNA to introduce nicks into the substrate. *E. coli* DNA polymerase I begins synthesis in a 5'→ 3' direction from those nicks. 5' [α-^{32}P]deoxyribonucleoside triphosphates are incorporated during synthesis, yielding radioactive probes for hybridization experiments. This technique was originally described by Rigby et al. (1977).

If the isotope is supplied in ethanol, then the appropriate aliquots must be dried in the reaction tube prior to the addition of the other components. However, most suppliers now make labeled compounds available as stabilized aqueous solutions that can be used directly. Nick translation kits are available from a number of vendors. These may suffice for most purposes.

PROCEDURE 49
DNA Purification with BND-Cellulose

Benzoyl-Naphthoyl-DEAE (BND)-Cellulose

1. Resuspend 15 g of BND-cellulose in a 250-ml graduated cylinder with column buffer (see below) containing 15% ethanol. Stir gently for 5 min, let settle for 10–20 min, then remove the supernatant containing unsettled BND-cellulose fibers (fines) from above the settled resin.

2. Add an equal volume of column buffer without ethanol. Stir for 5 min, then let settle for 10 min and remove the supernatant above the settled resin. Repeat twice.

3. Equilibrate with column buffer. The material can be stored at 4°C.

Column Preparation

Use a disposable polypropylene column. A useful column size is 0.5–0.7 cm wide by 4 cm long with a 10-ml column capacity.

1. Use 0.7–2.0 ml of resuspended BND-cellulose per column. A 1-cm packed bed will hold 1 mg of DNA or more.

2. Wash the bed with 10 ml of column buffer under low air pressure from any clean air source. Apply pressure until the meniscus is just above the bed.

3. Load the sample.

4. Wash with 5 ml of column buffer.

5. Elute with 0.5–1.5 ml of elution buffer (see below) into 1.5-ml microfuge tubes (0.5 ml in each).

6. Precipitate the DNA by adding 2 volumes of 95% ethanol (see Procedure 41, p. 180). Resuspend the DNA precipitates together in 50 µl of TE (10 mM Tris·HCl, 1 mM EDTA, pH 8.0).

Column buffer

    0.3 M NaCl
    0.01 M Tris·HCl, pH 7.4
    0.001 M EDTA

Elution buffer

    1.0 M NaCl
    15% ethanol

Notes

Obtain BND-cellulose from Accurate Chemical & Scientific Corp. BND-cellulose is a versitile ion-exchange resin for purification of nucleic acids. Recovery from the column of both high- and low-molecular-weight fragments is quantitative. Any losses usually occur at the ethanol precipitation step. The use of BND-cellulose for fractionation of nucleic acids has been described by Sedat et al. (1967).

PROCEDURE 50

Preparation of Mini-gel Filtration Columns

Prepare minicolumns in 1.5-ml microfuge tubes as follows:

1. Puncture the bottom of a microfuge tube with a 20-gauge needle.

2. Place a small amount of siliconized sand or glass beads in the bottom of this tube.

3. Place this tube in another 1.5-ml microfuge tube.

4. Fill the top tube with Sephadex G-50 (medium), Sepharose CL-6B, or Bio-Rad P60 (100–200 mesh) in a slurry (see Notes below) prepared in TE (10 mM Tris·HCl, 1 mM EDTA, pH 8.0).

5. Centrifuge the tubes at 1500 rpm for 20 sec in a table-top centrifuge fitted with a swinging-bucket rotor.

6. Inspect the prepared Sephadex columns. Save only those that pack evenly and have an unimpeded flow. Next replace the lower tube with a clean 1.5-ml microfuge tube. Reload the column with the same volume of TE as the sample size and centrifuge as above.

7. Inspect the lower tubes and discard those columns that deliver more or less than the loaded volume.

8. Repeat steps 6 and 7.

9. Replace and label the lower tube. Load samples on the minicolumn and centrifuge as in step 5.

10. The eluent in the bottom tube contains large-molecular-weight molecules.

Notes

This procedure can be used to remove unincorporated nucleotide triphosphates from nick-translation reactions (see Procedure 48, p. 199), for desalting DNA samples, and so forth.

Prepare resins according to the manufacturer's recommendations. The proper concentration of resin in the slurry can be estimated by letting the resin settle overnight. Buffer above the resin should be 1–2× the volume of the settled resin.

A convenient color test for minicolumn quality uses a solution of 0.25% bromophenol blue, 5% SDS, and 250 mM EDTA. Add 2 µl of this solution to your DNA sample and mix well. The dye is not excluded and gives a visual marker of material retained on the column.

PROCEDURE 51
Nuclease BAL-31 Digestion

1. Start with plasmid DNA banded in a CsCl equilibrium gradient (see Procedure 28, p. 144). Digest DNA with the appropriate restriction enzyme that will cut the plasmid once to form linear molecules. Extract the DNA with phenol and chloroform (see Procedure 40, p. 177). Precipitate the extracted DNA with ethanol (see Procedure 41, p. 180) and resuspend the DNA in sufficient NET buffer (5 mM NaCl, 10 mM Tris·HCl, pH 8.0, 0.1 mM EDTA) to give 0.1–1.0 µg/ml. Verify the complete conversion to linear molecules by running an agarose gel (see Procedure 43, p. 183). See Notes below.

2. The BAL-31 digestion involves treatment of 2 µg of linear plasmid DNA in a 50-µl reaction volume. Deletions generated by BAL-31 treatment will begin at each end of this linear molecule.
   Add the following to a microfuge tube in this order:

H₂O	up to 54 µl
10× BAL-31 buffer	5 µl (see below)
5 M NaCl	6 µl
DNA (from step 1)	2–20 µl (2 µg total)

3. Mix and incubate at 30°C for 1 min before adding enzyme.

4. To each of five microfuge tubes, add 3 µl of 0.25 M EDTA and 3 µl of 0.1 M EGTA. Keep these tubes on ice and on hand during the digestion. These tubes contain the STOP buffer.

5. Remove 10 µl of the mixture from step 3 into STOP buffer. This is the control sample, untreated with BAL-31.

6. Add 1 µl (0.5 units) of BAL-31 nuclease to the tube prepared in step 3; mix and incubate at 30°C. (1 unit equals the amount required to release 1 µg of acid-soluble nucleotides from heat-denatured calf thymus DNA at 650 µg/ml in 1 min at 30°C.)

7. At 1-min time points, remove 12 μl of the reaction mixture and add to the tubes with STOP buffer prepared in step 4. Keep the samples on ice. At the completion of the reaction, either keep each time-point sample separate for gel analysis of the digestion reactions or combine the samples at this step (see Notes below).

8. To each sample, add 1 μl of 10 mg/ml boiled yeast tRNA. This acts as a carrier and aids in complete recovery of DNA samples. Precipitate with ethanol (see Procedure 41).

9. Resuspend DNA pellet(s) in 10–20 μl of NET buffer. The extent of nuclease digestion at individual time points can be determined by running aliquots on a 0.7% agarose gel (Fig. 21). If the individual reactions from step 6 were combined, then a 1–5-μl aliquot can be analyzed by gel electrophoresis and the remainder of the DNA can be treated as described in steps 10–12.

10. Use 1–5 μl of the sample for ligation in a total volume of 20 μl (see Procedure 31, p. 154).

11. Add 1 μl of T4 DNA ligase (10 units) and mix.

12. Incubate on ice for 6–18 hr and proceed to transform competent cells (see Procedure 37, p. 169).

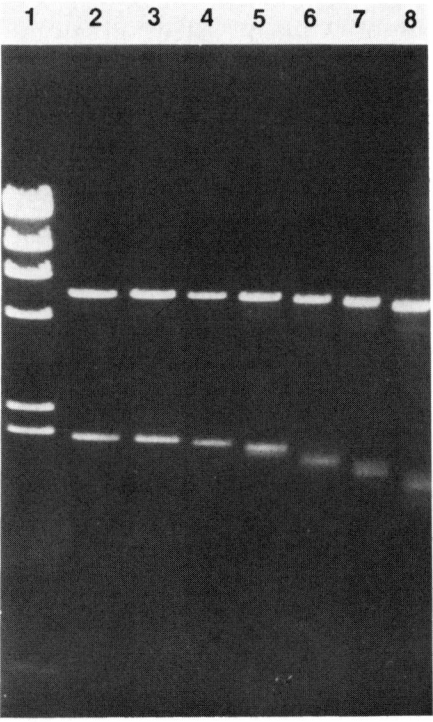

Figure 21
Titration of BAL-31 nuclease. Approximately 50 μg of a 7-kb plasmid were digested with an enzyme that cuts this DNA only once to give linear molecules. This DNA was treated with BAL-31 as described. Following treatment, each DNA sample was digested with a second restriction enzyme that cleaves at a second site, yielding two DNA bands of ~5.2 kb and ~1.8 kb. (Lane 1) λDNA marker cleaved with HindIII; (lane 2) no BAL-31; (lane 3) 2 min; (lane 4) 5 min; (lane 5) 10 min; (lane 6) 15 min; (lane 7) 20 min; (lane 8) 30 min. Note that the BAL-31 digestion is more evident with a smaller DNA fragment. (Gel provided by H. George, Molecular Genetics).

Notes

To eliminate the need for the extraction and precipitation described in step 1, digest the plasmid according to Procedure 31 (p. 154). After digestion, heat the sample to 70°C for 5 min to inactivate the restriction enzyme. The total volume is then adjusted to 55 μl by using water and buffer components to yield a final solution of:

    12 mM $CaCl_2$
    12 mM $MgCl_2$
    20 mM Tris·HCl, pH 8.0
     1 mM EDTA
  600 mM NaCl

Samples treated in this way can be subjected to BAL-31 digestion (step 3). This alternative method is recommended when treating small amounts of DNA to avoid the losses that usually accompany extraction and precipitation.

To verify that the BAL-31 digestion is working, run two consecutive experiments, using the same DNA and enzyme conditions. Run the individual time-point samples from the first experiment on a gel and look for the decrease in size of the linear plasmid band. This is an indication of the extent of BAL-31 digestion.

If the reaction seems slow, increase the incubation temperature. BAL-31 has an optimum temperature of 60°C.

If the reaction is too fast, lower the incubation temperature or decrease the salt concentration.

BAL-31 will probably not remove exactly the same number of nucleotides from each end. In fact, assymetry is often observed.

Do not use too much enzyme. Some workers report that more than 10 units/ml tend to leave many unligatable ends. It is much better to use low-enzyme concentrations with longer incubation times.

Higher efficiency of ligation may be obtained by repairing the BAL-31-digested DNA with DNA polymerase J.

A synthetic oligonucleotide linker containing useful restriction enzyme cleavage sites can be added at the site of BAl-31 digestion. The phosphorylated linker can be added in step 9.

This procedure was developed by H. George of Molecular Genetics, Inc.

10× BAL-31 buffer

  120 mM $CaCl_2$
  120 mM $MgCl_2$
  200 mM Tris·HCl, pH 8.0
   10 mM EDTA

PROCEDURE 52

Preparation of SDS Protein Extracts

DAY 1

1. Prepare a culture tube with 5 ml of L medium.

2. Inoculate with a single colony of the strain to be analyzed and shake or rotate overnight at 30°C.

DAY 2

1. Inoculate 0.2 ml of cells in 10 ml of L medium and shake or rotate at 37°C.

2. Grow the culture to late-log phase. Measure the cell density (OD_{600}) in a spectrophotometer. Calculate the appropriate volume of cells according to the following formula:

$$X \text{ ml} = \frac{4.8}{OD_{600}}$$

   (i.e., if OD_{600} = 1.0, use 4.8 ml of cells.)

3. Centrifuge the appropriate volume of cells at $1500g$ for 10 min.

4. Resuspend the cell pellet in 0.35 ml of 2× loading buffer (LB; see Procedure 53, p. 209). Transfer to a 1.5-ml microfuge tube.

5. Boil the samples in a water bath for 5 min.

6. Centrifuge the samples in a microfuge (~$12,000g$) for 5 min.

7. Load 30–35-μl aliquots on a polyacrylamide gel (see Procedure 53).

PROCEDURE 53
Electrophoresis of Proteins

Stocks

Acrylamide stock

    300 g acrylamide
      8 g bisacrylamide
    H_2O up to 1 liter

Filter through a 0.45-μm filter. Store acrylamide stock at 4°C.

10% SDS

    10 g sodium dodecyl sulfate (SDS)
    H_2O up to 100 ml

Store at room temperature.

4× Lower buffer (1.5 M Tris·HCl, pH 8.8, 0.4% SDS)

    181.7 g Tris base
     40 ml 10% SDS
    H_2O up to 1 liter

Adjust the pH to 8.8 with HCl. Add 1 ml of TEMED (N,N,N',N'-tetramethyl-ethylenediamine).

4× Upper buffer (0.5 M Tris·HCl, pH 6.8, 0.4% SDS)

    60.6 g Tris base
    40 ml 10% SDS
    H_2O up to 1 liter

Adjust the pH to 6.8 with HCl. Add 2 ml of TEMED.

0.1% BpB

    10 mg bromophenol blue
    10 ml H_2O

Store at room temperature.

2× Sample buffer (SB)

    12.5 ml 4× upper buffer
    20.0 ml glycerol
    H_2O up to 60 ml

2× Loading buffer (LB)

    0.5 ml β-mercaptoethanol
    0.25 ml 0.1% BpB
    4.0 ml 10% SDS
    5.3 ml 2× SB

4× Running buffer stock (RBS)

    60 g Tris base
    288 g glycine
    H_2O up to 5 liter

1× Running buffer (1× RB)

    1 liter 4× RB
    3 liter H_2O
    40 ml 10% SDS

Stain

    125   ml isopropanol
    50   ml acetic acid
    325   ml H_2O
    1.25 g Coomassie brilliant blue (R250)

Destain

    1.0 liter methanol
    1.4 liters acetic acid
    H_2O up to 20 liters

Gel

The following procedure can be used to cast a gel 20 cm × 17 cm × 1.5 mm.

1. Using the amounts given in Table 2, cast the lower gel in an appropriate apparatus. Allow room for a 3-cm stacking gel. With a pasteur pipette, carefully layer 0.1% SDS on the surface of the gel. Polymerization will require 1–2 hr.

Table 2
Lower gel

Stock	Acrylamide concentration (final)		
	8%	10%	12%
4× Lower buffer	10 ml	10 ml	10 ml
Acrylamide stock	10.6 ml	13.3 ml	16.0 ml
H_2O	19.2 ml	16.5 ml	13.8 ml
10% Ammonium persulfate (freshly prepared)	0.2 ml	0.2 ml	0.2 ml

Stacking gel (3%)

    2.5 ml 4× upper buffer
    1.0 ml acrylamide stock
    6.4 ml H_2O
    0.2 ml 10% ammonium persulfate (prepared fresh)

    First remove the SDS from the top of the lower gel by rinsing with H_2O. After pouring the stacking gel, clamp in comb and allow to polymerize (1–2 hr).

3. Remove comb, place the gel in the apparatus, and fill the upper and lower buffer chambers with 1× RB. Clean the gel slots with running buffer.

4. Attach the leads to the power supply and connect the *positive* pole to the lower buffer chamber.

5. Load your sample (10–70 µl) and run electrophoresis at 5–10 V/cm.

6. Run the electrophoresis until the blue-dye front reaches the bottom of the gel or has just run off. Remove the gel from the glass plates.

7. Stain the gel for 30–60 min. Save the stain; it can be reused several times.

8. Destain the gel in several changes of destain. Addition of ion-exchange resin beads (Dowex 1XB, 20–50 mesh) or plastic foam flask stoppers facilitates the destaining. Place the destaining gel on a slow platform shaker. Complete destaining may take 8–16 hr.

9. The destained gel can be dryed for a permanent record or for autoradiography. Wash the gel in water, place on Whatman 3MM paper or its equivalent, and cover with plastic wrap. Dry on a gel drier for 1–2 hr.

Note

This procedure is adapted from Laemmli (1970).

PROCEDURE 54
Maxicells: Plasmid Gene Expression

DAY 1

1. Prepare a culture tube with 5 ml of M63 medium containing 1% casamino acids, 0.4% glucose, and an appropriate antibiotic.

2. Inoculate with a single colony of SE5000 carrying the plasmid to be analyzed and shake or rotate overnight at 30°C.

DAY 2

1. Inoculate 0.2 ml of the cells into 11 ml of the same medium and shake or rotate at 37°C until the culture reaches an OD_{600} of 0.7.

2. Irradiate the cells in a large, uncovered petri dish that is rotating on a turntable in the dark with a UV light source at the appropriate distance above the dish to give a dose of 300 erg/mm^2. Irradiate with occasional mixing.

3. Pipette 10 ml of the irradiated cells into an amber or foil-wrapped test tube.

4. Shake or rotate for 30 min at 37°C and then add 40 µl of a 50 mg/ml stock of cycloserine. Continue incubation overnight.

DAY 3

1. Spread 0.1 ml of cells on an L plate. Incubate the plate at 37°C overnight. (There should be few or no survivors.)

2. Centrifuge the cells at 1500g for 10 min. Resuspend the cell pellet in 10 ml of M63 medium and recentrifuge.

3. Repeat step 2.

4. Resuspend the cell pellet in 2 ml of M63 medium containing 0.4% glucose.

5. Incubate at 37°C for 1 hr with aeration.

6. Add [^{35}S]methionine to 25–100 µCi/ml.

7. Incubate from 45 sec to 30 min, depending on the experiment. Most of the label is incorporated in the first few minutes.

8. Centrifuge the cells at 1500g for 10 min. Remove and discard the radioactive supernatant.

9. Resuspend the cell pellet in 0.5 ml of M63 medium and transfer to a 1.5-ml microfuge tube.

10. Spin for 5 min in a microfuge (~12,000g). Remove and discard the radioactive supernatant.

11. Repeat steps 9 and 10.

12. Resuspend the final cell pellet in 0.1 ml of 2× loading buffer (LB; see Procedure 53, p. 209). Boil for 5 min in a water bath and store at −20°C. Run electrophoresis with a 5–10-µl aliquot on an SDS-polyacrylamide gel (see Procedure 53).

13. Follow the autoradiography procedure described in Procedure 46 (p. 191).

Notes

This is a modification of the procedure described by Sancar et al. (1979).

APPENDIXES

APPENDIX A
Media and Standard Solutions

Unless otherwise stated, media should be sterilized by autoclaving for 20 min at 121°C, (15 lb/in.^{2}). For quantities of media greater than 2 liters, longer sterilization times are required.

Rich Media

L medium (LB, Luria-Bertani)

Per liter:

Bacto-tryptone	10 g
Bacto-yeast	5 g
NaCl	5 g

For plates:

Bacto-agar	15 g

For top agar:

Bacto-agar	7 g

Other common variations:

1. L citrate broth: Add 2 ml of 1 M sodium citrate per 100 ml of L medium (20 mM final concentration).

2. L citrate tetracycline agar: Spread an L plate containing tetracycline (25 µg/ml) with 0.1 ml of 1 M sodium citrate (4 mM final concentration).

TB medium (tryptone broth, λ medium)

Per liter:

Bacto-tryptone	10 g
NaCl	5 g

For plates:

Bacto-agar	11 g

For top agar:

Bacto-agar	7 g

After autoclaving, add 10 ml of 1 M $MgSO_4$.

Indicator Agar

Tetrazolium agar

Bacto-antibiotic medium 2	25.5 g
2,3,5 triphenyl-2H tetrazolium chloride	50 mg (5 ml of a 1% solution)
H_2O	950 ml

Autoclave for 30 min. Afterwards add 50 ml of 20% sugar solution.

MacConkey agar

Bacto-MacConkey agar base	40 g
H_2O	950 ml

After autoclaving, add 50 ml of 20% sugar solution. For MacConkey top agar, use 20 g of agar base per liter, omitting the sugar.

TTC agar[1]

Bacto-tryptone	10 g
Bacto-agar	10 g
NaCl	5 g
2,3,5 triphenyl-2H tetrazolium chloride	25 mg (2.5 ml of a 1% solution)
H_2O	950 ml

After autoclaving, add 50 ml of 20% sugar solution and 2 ml of thiamine (50 mg/ml).

---

[1] Galactose-TTC agar is used for the red-plaque test (Procedure 5, p. 97).

Minimal Agar

M63 medium

Solution A:

    Bacto-agar   15 g
    H_2O 500 ml

Solution B:

    10× M63 stock   100 ml
    H_2O 400 ml

After autoclaving separately, mix solutions A and B, then add 10 ml of 20% sugar solution, 2 ml of thiamine (50 mg/ml), and 1 ml of 1 M $MgSO_4$.

10× M63 stock

Per liter:

    KH_2PO_4 30 g
    K_2HPO_4 70 g
    $(NH_4)_2SO_4$ 20 g
    $FeSO_4$ 5 mg (5 ml of 1 mg/ml solution)

Add $FeSO_4$ after salts are dissolved at final volume. It is not necessary to sterilize this solution. For liquid M63 (1×), dilute 10× stock and autoclave in 100-ml aliquots. Before use, add:

    0.1 ml 1 M $MgSO_4$
    0.2 ml thiamine (50 mg/ml)
    1.0 ml 20% sugar

F top agar

    Bacto-agar   8 g
    NaCl   8 g
    H_2O 1 liter

Water top agar (W agar)[2]

    Bacto-agar   7 g
    H_2O 1 liter

Agarose top agar[3]

    Agarose   7 g
    H_2O 1 liter

---

[2] Water top agar is used in Procedure 47 (p. 196).
[3] Agarose top agar is used in Procedure 2 (p. 91).

Stab Agar

Per liter:

Bacto-nutrient broth	10 g
NaCl	8 g
Bacto-agar	6 g

Heat to dissolve and dispense into Wheaton vials prior to autoclaving (see Appendix D, p. 231).

Antibiotic Agars

After autoclaving (with agar at ~50°C), add an appropriate volume of antibiotic stock. Stocks should be sterilized by filtration.

1. For tetracycline hydrochloride, add 2.5 ml of a 10 mg/ml stock per liter (25 µg/ml final). For tetrazolium and MacConkey indicator media, use 1/3 the amount of tetracycline (8 µl/ml final).

2. For kanamycin sulfate, add 1 ml of a 25 mg/ml stock per liter (25 µg/ml final).

3. For streptomycin sulfate, add 1 ml of a 125 mg/ml stock per liter (125 µg/ml final).

4. For chloramphenicol, add 10 ml of a 2.5 mg/ml stock per liter (25 µg/ml final).

5. For ampicillin (sodium salt), add 1 ml of a 125 mg/ml stock per liter (125 µg/ml final). For Mud lysogens, use 1/5 the amount of ampicillin (25 µg/ml final).

An alternative method is to spread the antibiotic from a stock solution on to the desired plate. Spread 0.2 ml of a 100× stock per plate. Because it is difficult to ensure uniform diffusion, it is preferable to add antibiotics prior to pouring plates.

Standard Stock Solutions for Media Preparation

Unless otherwise stated, these solutions should be filter sterilized.

1 M $MgSO_4$

1 M Sodium citrate

1 M $CaCl_2$

Thiamine (50 mg/ml); store at 4°C

2,3,5 Triphenyl-2H tetrazolium chloride (10 mg/ml)

$FeSO_4 \cdot 7H_2O$ (1 mg/ml)

Sugars, generally 20% w/v (autoclave in 100-ml aliquots for 15 min)

Amino acids, generally 10 mg/ml

Xgal (5-bromo-4-chloro-3-indolyl-ß-D-galactoside) in *N-N*-dimethylformamide (20 mg/ml). Sterilization not required—*do not filter.* Store at 4°C

Usually amino acids and Xgal are spread on plates. Spread 0.1 ml of the above stock solutions.

Buffers

λ-dil (TMG buffer)

Per liter:

Tris base	1.21 g
$MgSO_4 \cdot 7H_2O$	1.20 g
Gelatin	0.10 g

Adjust pH to 7.4 with HCl. Sterilize by autoclaving.

TE buffers

Stock solutions:

1 M Tris·HCl:

Tris base	121 g
H_2O	1 liter

Adjust to desired pH with HCl.

250 mM EDTA:

$Na_2 \cdot EDTA \cdot 2H_2O$	93 g
H_2O	1 liter

Sterilize by autoclaving. Store stock solutions at 4°C. Before use, dilute stocks using autoclaved H_2O to obtain desired final concentration. One of the most common TE buffers for DNA biochemistry is 10 mM Tris · HCl, pH 8.0, 1 mM EDTA.

SSC buffer

20× stock per liter:

NaCl	175.4 g
Sodium citrate	77.4 g

Adjust pH to 7.0 with NaOH. Store at 4°C.

DNase and RNase stocks

Stock solutions of RNase A are prepared as follows: Dissolve RNase A in 25 mM Tris·HCl, pH 7.4, at 5 mg/ml. Boil for 5 min to destroy contaminating enzyme activities (particularly DNases). Add an equal volume of 40% glycerol and 0.5 volume of 0.5 M NaCl. The final stock (2 mg/ml, 10 mM Tris·HCl, pH 7.4, 0.1 M NaCl) is stable and can be stored in aliquots at −20°C.

Purchase RNase-free DNase I in 5-mg lots from Worthington Biochemicals (cat. no. DPFF, 2000–2600 units/mg). Resuspend the entire lot in 50% glycerol, 10 mM Tris·HCl, pH 7.5. Store at −20°C in 100-μl aliquots.

APPENDIX B
Phenotypes and Genotypes of *Escherichia coli*

Strains used in bacterial genetics are described by genotype. The following discussion should aid you in writing and interpreting *E. coli* genotypes.

The first symbol in the genotype describes the mating type of the strain (e.g., Hfr, F′, F⁺, or F⁻). If you are not familiar with bacterial mating, you should consult an introductory genetics text or Miller (1972). The origin of transfer of common Hfr strains and the bacterial genes carried by common episomes are described in Bachmann and Low (1980) and Low (1972). In addition, a useful pedigree of common *E. coli* K-12 strains has been compiled (Bachmann 1972).

Most of the other information contained in the genotype refers to specific mutations. A mutation is an inheritable change in the chromosome of an organism. The altered genotype may or may not confer a recognizable phenotype. To read the genotype (see Bacterial Strains, p. xi) and predict the phenotypes of any given *E. coli* strain, it is important to recognize some broad mutant categories.

E. coli is a prototrophic organism. As such, it will grow on synthetic media containing only the essential inorganic salts and a suitable carbon source. Auxotrophs are mutant strains with a defect in an anabolic pathway and, as a result, are unable to grow on this defined medium. Such mutants require additional organic compounds, such as vitamins, amino acids, or nucleosides, for growth. Since these compounds are generally present in rich media (L medium), no additions are needed for growth.

A second common class of mutants comprises those strains unable to utilize a particular carbon source. They carry mutations that inactivate a catabolic pathway. On any media containing an alternate carbon source, these mutants will grow normally.

A third category of *E. coli* mutants is composed of those that harbor defects in essential genes, e.g., those whose products are involved in protein synthesis and DNA replication. To permit cell survival, mutations that remove an essential function must be conditional; i.e., under a particular condition (permissive), they must be phenotypically silent. There are two common types of conditional mutations, those leading to an altered gene product that is functional only at certain growth temperatures and those requiring a second mutation that phenotypically suppresses the original lesion.

There are many kinds of suppressor mutations. Some act at the informational level, e.g., amber suppressors (an alteration in the anticodon loop of a tRNA that results in the insertion of an amino acid at a chain termination codon). Others can act by causing production of a second mutant product that interacts directly with the original mutant component, restoring the wild-type (nonmutant) phenotype. Unlike amber suppressors, these latter suppressors will suppress only certain mutations in a particular gene; i.e., they possess allele specificity.

The final common type of mutant carries mutations that confer resistance to toxic agents such as bacteriophages, colicins, or antibiotics. Although a number of different cellular components can be affected by these mutations, they often confer no easily recognizable phenotype in the absence of the toxic agent.

The following are some brief guidelines to standard *E. coli* genetic nomenclature (Bachmann and Low 1980; Bachmann 1983). Mutant loci are usually described by a three-letter designation written in lower-case, italicized letters (e.g., *lac*). The abbreviation is usually based on a recognizable phenotype (e.g., ability to grow on lactose). Individual genes that affect the phenotype are described by an italicized capital letter immediately following the three-letter designation (e.g., *lacZ*). Letters or numbers following the capital letter refer to specific mutations (allele numbers; e.g., *lacZU118*). Generally, only mutant genes are listed in a genotype. For example, a *lac* strain cannot grow on lactose. Occasionally, a superscript + is used to emphasize the fact that a particular gene is nonmutant. In any case, if a gene is not listed, it is assumed to be the wild type.

Phenotypes are usually abbreviated by using the same three-letter designation as the genotype. However, the phenotype is not italicized and the first letter is capitalized (e.g., Lac). The superscripts + and − are used to denote the wild-type or mutant, respectively (e.g., Lac$^+$, Mal$^-$). Exceptions to this rule include mutations that confer resistance to antibiotics and bacteriophages. In the former case, a standard, two-letter code has been adapted for most every antibiotic (e.g., Ap, Tc); in the latter case, the name of the phage is usually employed (e.g., λ). The superscripts s and r denote sensitivity or resistance, respectively (e.g., Ap^s, λ^r).

The Greek letter Φ is used to designate fusions. Fusion components are written in the order of transcription. Protein fusions are designated Φ(*ompR'-'lacZ*)*hyb*; operon fusions are designated Φ(*ara'-lacZ*$^+$). In these examples, the prime indicates that the DNA for the gene is deleted at the position of the prime (e.g., *'lacZ* denotes a deletion in the 5' region of the gene). Numbers appearing after the fusion designation are allele numbers.

Several common symbols are used to designate certain types of mutations. The Greek letter Δ indicates a deletion [e.g., Δ(*lac*)*U169*]. The genes enclosed in parentheses have been deleted. Again, the *U169* in this example refers to an allele. The symbol :: represents an insertion. The format is target gene::insertion element (e.g., *lacY*::Tn9). In this case, allele numbers follow the target gene.

Because it is possible to isolate insertions that are near a particular gene but not within any known gene, the following system was developed. Using a three-letter code, the site of the insertion is named according to the position measured in minutes on the *E. coli* linkage map. The first letter is always z, indicating an insertion in no known gene, the second letter indicates the 10-min interval, and the third letter identifies the 1-min interval (e.g., *zab*::Tn*10* is a Tn*10* transposon inserted between 1 and 2 min on the chromosome).

Finally, strains lysogenic for a particular bacteriophage are designated with the name of the phage in parentheses at the end of the genotype. A similar notation is used for strains carrying plasmids. The plasmid name, a lower-case p followed by a set of initials and a number, appears in parentheses at the end of the genotype [e.g., MC1000(pMLB1034)].

We realize that reading a genotype can be complicated. If you are unsure, consult the current *E. coli* linkage map (Bachmann 1983).

APPENDIX C
Transposable Genetic Elements

Transposable elements (transposons) are defined sequences of DNA capable of transposing from one replicon to another by a mechanism that is independent of homologous recombination. Insertion of a transposable element into a replicon such as the bacterial chromosome can occur at many different sites, and, for our purposes, we can consider this event to be essentially random. However, with respect to the transposon itself, insertion is specific in that the linear structure of the element is never permuted.

Generally speaking, transposable elements can be divided into four classes based on size, structure, and gene products specified. Class I elements are the smallest (750–1500 bp). The product(s) specified by these elements participate directly in the transposition reaction. Often these elements are called insertion sequences (IS). Some common examples are: IS*1*, IS*10*, and IS*50* (see Table 3).

Class II elements, in addition to specifying products that participate in the transposition reaction, also carry other determinants, such as antibiotic resistance. These transposons (Tn, or drug-resistance elements) are composed of a gene(s) for the extra determinant flanked by IS elements in either direct or indirect repeated form. (Originally, the term transposon was reserved for elements carrying extra determinants; however, this definition has fallen from favor.) In this configuration, the IS elements can transpose individually or in concert; in the latter case, the flanked marker is transposed as well. Often, one of the flanking IS elements is defective and cannot transpose by itself. By convention, the nonfunctional, defective IS element is termed left (L), and the functional element is called right (R) (e.g., IS*10*R). The transposons used or referred to in this manual are summarized in Table 3.

Class III elements resemble insertion sequences except that they carry an extra determinant and consequently are somewhat larger. However, unlike transposons such as Tn*10*, they do not contain flanking IS elements. The most common members of this class are closely related to Tn*3*.

Class IV transposable elements are the largest of the known prokaryotic transposons. They are bacteriophages such as Mu.

Since their discovery, transposable elements have generated considerable interest. This has been heightened by the realization that eukaryotic cells also contain similar elements. Current research in this area is fo-

Table 3
Transposons Commonly Employed in Bacterial Genetics

Transposon	Length (kb)	Associated determinant	Flanking IS element	Size of IS element (bp)	Orientation of IS element
Tn3	4.9	Ap^r	NA	NA	NA
Tn5	5.7	Km^r	IS50	1534	inverted
Tn9	2.5	Cm^r	IS1	768	direct
Tn10	9.3	Tc^r	IS10	1329	inverted
Mu	37.7	phage genes	NA	NA	NA

(NA) Not applicable; (Ap) ampicillin; (Km) kanamycin; (Cm) chloramphenicol; (Tc) tetracycline.

cused on understanding the molecular mechanism(s) of transposition. A discussion of this research is beyond the scope of this manual. Rather, we will focus on transposons as an important tool for bacterial genetics. For a detailed description of transposons and transposition, see the recent review by Kleckner (1981).

All transposons are mutagens that cause defined mutations with the following properties: (1) insertion of a transposon into a gene will disrupt the linear continuity of that gene and destroy its function, resulting in a null phenotype (i.e., a knock-out mutation); (2) such an insertion in an operon will be strongly polar; (3) it will revert by precise excision; and (4) it will behave as a point mutation in a genetic cross. Moreover, unlike many other mutagens, transposons cause mutations at high frequency after low-level mutagenesis. Thus, one need not be overly concerned about "multiple hits" during mutagenesis.

Drug-resistance elements are particularly useful because they carry a selectable marker. Consequently, all mutations caused by these elements will exhibit two phenotypes; the null phenotype caused by the insertion mutation and the drug-resistance phenotype. Since both phenotypes are the result of the same insertion event, they will be completely linked in all subsequent genetic experiments. This is useful for strain construction. For example, suppose that we wish to make a particular strain Mal$^-$. Normally this could not be done by generalized transduction or mating because a Mal$^-$ phenotype cannot be selected. However, with *mal*::Tn10 this is no longer a problem. We simply select for resistance to tetracycline (Tc^r). All transductants or exconjugates that are Tc^r will be Mal$^-$ (see Experiment 9, p. 59).

Strain construction is further enhanced with drug-resistance elements because chromosomal insertions can be isolated even if the insertion does not destroy a gene and cause a recognizable phenotype. This permits the isolation of insertions that are near but not within a particular gene (see Experiment 8, p. 53). Accordingly, any allele of any gene can be crossed genetically to another strain by selecting drug-resistance and scoring for the allele of interest. For example, suppose that we wish to transfer a temperature-sensitive, lethal mutation from one strain to another. (Note that this mutation causes a phenotype that is unlikely to be null. Thus, one could not, as in the case of Mal⁻, obtain this phenotype by insertion. In all probability, an insertion in this particular gene would be lethal under all conditions.) Here again, conventional methods are complicated because one cannot select lethality. However, by inserting a drug-resistance element near this mutation, it can be moved to any other strain by selecting drug resistance under permissive conditions and then scoring for the temperature-sensitive phenotype.

A perhaps more subtle property of transposons is that they provide a region of genetic homology that can be inserted into different nonhomologous replicons. This portable homology provides the geneticist with a technical power that approaches recombinant DNA methodology (see Experiment 1, p. 7). The specific applications that have been developed based on this property are far too numerous to mention. However, remember that the essence of recombinant DNA methodology is that it provides a means, using various purified enzymes in vitro, for forming a novel DNA joint. The enzymes of transposition do the same thing in vivo, and because of the power of genetic selection and homologous recombination, such reactions can be controlled.

A final property of transposons that can be exploited is that these elements promote several types of chromosomal rearrangements, e.g., inversions or deletions. Although useful under certain circumstances, this property is actually a mixed blessing. All bacterial strains contain transposable elements. If you purposefully add to this by constructing strains that carry one or more drug-resistance elements, you are increasing the potential for transposon-mediated events. Keep in mind that environmental stress (any selection procedure) may well stimulate these effects (Arber et al. 1981). As a consequence, the literature is full of seemingly unexplainable results and numerous incorrect conclusions. The cliché "caveat emptor" is especially applicable here.

The summary of experimental applications of transposons presented here is by no means exhaustive. For a more detailed discussion, see Kleckner et al. (1977).

Researchers exposed to transposons for the first time are often confused by the plethora of known examples; how do you know which one to use for any given experiment? There is no perfect answer; however, certain transposons have characteristic properties that are useful under particular circumstances. In this manual, five elements, or portions or derivatives of these elements, are used. The important characteristics of each are described below (Kleckner 1981). Before beginning, however, an additional property of transposons in general must be mentioned.

All transposons come equipped with a regulatory mechanism to control transposition frequency. Teleologically, this makes sense, because any Tn element that transposed in a totally unrestrained fashion would quickly kill the host cell. Any element so inclined would soon become extinct. As you may have guessed, the regulatory mechanisms employed by different transposons differ widely, and some are more effective than others.

Mu

This bacteriophage transposes at relatively high frequencies. Accordingly, it is not difficult to isolate Mu insertions in any given gene. Its limitations are three. First, it is very large (38 kb). Second, although it carries many genes, none of them is easily selected. [This limitation does not apply to the Mud(*lac*, Ap) phage]. Finally, the gene products that promote transposition are required for lytic growth and, like all of the genes, are controlled by a repressor that acts in a manner not unlike the λ repressor (see Appendix F, p. 236). As such, transposition is stimulated when the genome is introduced into virgin cells that do not contain Mu repressor (zygotic induction). This complicates genetic transfer of Mu or Mu*d* insertions from one strain to another. When selection is for a marker present on the phage, it is not uncommon to find that the Mu present in the recipient cell is not at the same place as was the Mu in the donor cell (see Experiment 1; Silhavy and Beckwith 1983). Moreover, care must be taken to ensure that the Mu does not transpose in the donor cell to an additional location in the chromosome, forming a multiple lysogen. (These limitations do not apply to λ*plac*Mu; see Appendix L, p. 261).

Tn5

In many respects Tn5 resembles Mu. It transposes at high frequency, and thus isertion mutations are relatively easy to isolate. Transposition is stimulated in virgin cells (the mechanism is, however, much different [Isberg et al. 1982]), and the element can often be found in more than one location in a single cell. Unlike, Mu, Tn5 is small and confers a readily selectable phenotype (Km^r). All in all, Tn5 is very useful. It does require, however, a certain amount of special care.

Tn3

The determinant carried by Tn*3* provides a strong selection. This element is of limited use because it appears to favor greatly insertion into certain replicons such as plasmids rather than the chromosome (Kretschmer and Cohen 1979). Although we do not use Tn*3* in this manual, we do use the drug-resistance element of the transposon. This gene specifies a β-lactamase and is present in the Mu*d* phages and plasmids derived from pBR322.

Tn9

This element provides a strong selection (Cm^r); however, it transposes at low frequency. Moreover, since the gene specifying drug resistance is flanked by IS*1* in a direct repeat, the selective marker can be lost by homologous recombination.

Tn10

This element transposes at low but reasonable frequencies. It confers an easily selectable phenotype (Tc^r), and Tn*10* insertions are stable. They can be transferred from strain to strain by generalized transduction or mating without difficulty. A further advantage is the availability of media that permit selection for loss of drug resistance (Tc^s; Bochner et al. 1980; Maloy and Nunn 1981). Although Tn*10* is large, it is probably the transposon of choice for bacterial genetics.

APPENDIX D
Notes on Growth and Storage of Escherichia coli

Growth of bacteria like *E. coli* in liquid is characterized by three stages. An initial lag phase, followed by a period of exponential growth that slows as cells enter a stationary phase. In the laboratory this growth can be followed by measuring viable cells per culture volume or more simply by measuring the turbidity of the culture. This turbidity is often quantitated by determining the optical density at 600 nm. Although the number of viable cells and the OD_{600} are directly proportional, the conversion factor may vary slightly, depending on the particular strain or growth condition. With practice it is easy to estimate the density of a 5-ml culture by eye with sufficient accuracy for most experiments.

E. coli is a facultative anaerobe and therefore will grow under anaerobic conditions. However, to obtain the maximal growth rate in any given media, vigorous aeration is required. This is usually achieved by mechanical agitation (rolling tubes and shaking flasks). It is important to note that this method works only if the culture volume is less than 25% of the container size. We cannot stress enough the need to *keep culture growth conditions constant*. Genetic analysis can be complicated enough without adding the many variables involved in anaerobic and aerobic growth. Cells growing in tightly capped tubes that are slowly shaking in a water bath are in a different physiological state than those in a vigorously shaking flask.

STORAGE OF *E. coli* CULTURES

Numerous methods have been described for both short- and long-term storage of *E. coli*. We recommend the following.

Short-term Storage

It is unwise to store *E. coli* as colonies on L plates at 4°C for periods of greater than 10 days. Such storage will reduce cell viability. On indicator media, the problem is compounded by the presence of toxic indicator dyes. Moreover, indicator media may offer a selective advantage for certain mutants. For strains that are used frequently, best results can be obtained by storage at 4°C in MgSO$_4$ suspension. A fresh overnight culture is centrifuged (3500 rpm for 10 min), the supernatant is discarded, and the pellet is resuspended in 0.5 volume of 10 mM MgSO$_4$. Under these conditions, cells remain viable for up to 1 month.

Long-term Storage

One of the most practical and reliable methods for long-term storage of *E. coli* cells is freezing at −70°C. This is done using screw-capped, plastic freezing vials (2-ml pro-vials or Nunc tubes) or 1-dram Wheaton glass vials.[1]

In one vial add:

0.4 ml overnight culture
0.6 ml 20% glycerol

Mix. Label.

Freeze two vials per culture. One is the working stock. The other is permanent, to be used in emergencies only. Storage in two separate freezers is a good idea.

Vials can be frozen and thawed many times with little loss of viability. Even so, it is best to scrape off a little frozen culture and streak it out (use a strong loop) rather than thaw each vial completely.

An alternative method, often used in conjunction with freezing at −70°C, is the storage of *E. coli* in stabs at room temperature. These are prepared by inoculating stab agar medium (see Appendix A, p. 217) in 1-dram Wheaton vials. Although reliable, this method is not recommended for Hfr or F' strains, strains carrying transposable elements such as Mu, Mud(*lac*, Ap), or Tn5, or strains containing plasmids such as pBR322 and its derivatives.

---

[1] Glass vials for storage at −70°C or for preparation of stabs can be obtained from Wheaton Industries. The dimensions of the 1-dram vial (cat. no. 224882) are 15×45 mm, and it has a capacity of about 4 ml. To prevent evaporation, use rubber-lined caps (as opposed to Teflon- or paperlined). **Note:** Do not use glass vials in a liquid nitrogen freezer!

APPENDIX E
Notes on Growth and Storage of λ

The best cells for lytic growth of λ are those in log phase. Cells in stationary phase are poor hosts for lytic growth. This is easily demonstrated by observing λ plaques: They are small and of finite size. This is so because when cells stop growing, so does λ. Phage λ lysogeny is enhanced in stationary-phase cells. A general rule of thumb is to use log-phase cells for lytic growth, and to use stationary-phase cells to make lysogens. It is important to know that λ requires Mg^{++} for stability and adsorption to *E. coli*.

It is convenient to use overnight cells for most phage work. However, it is inadvisable to have the cells reach stationary phase and then sit in their own excrement for hours before use. To have the overnight cells as close to log phase as possible in the morning, it is best to incubate cells at 30°C to slow their growth. In addition, we use a less rich medium, TB medium, rather than rich media containing yeast extract (L medium). In TB medium, cells stop growing at an OD_{600} of about 0.5–0.6. In L medium, cells continue to grow to an OD_{600} of 2.0 or more. L medium overnight cultures are inferior to TB overnight cultures for λ work. A common observation is that overnight cultures in L medium give lower bursts of λ in single infections. If such cells are diluted 1/50 and grown to an OD_{600} of about 0.2, they are fine for λ. Similarly, L plates give smaller, less-uniform plaques than do TB plates. Consequently, for most λ work (except plaque blotting, Experiment 6, p. 43), use TB plates.

A seeming contradiction is the use of L medium for liquid lysates of λ. Remember that as long as cells are in log phase, λ grows well. Rich media ensures rapid cell growth and also good phage yield. It is overnight growth in L medium that yields poor cells for λ work.

Maltose and Bacteriophage λ

Bacteriophage λ adsorbs, via its tail fiber, to an *E. coli* outer membrane protein, LamB (the product of the *lamB* gene; see Appendix N, p. 283). Strains that lack this protein are λ resistant ($λ^r$) since the phage cannot adsorb to them. The *lamB* gene is part of the maltose regulon and is therefore inducible by maltose. However, even in the absence of maltose, some *lamB* product is made. This means that λ will adsorb to cells not grown in maltose (e.g., in glucose). When cells are grown in 0.2% maltose, the *lamB* product is induced and many more λ receptors are present on the cell surface. This results in several thousand receptors per cell, as opposed to a few hundred per cell in noninduced cells. The first stage of λ adsorption is a purely physical interaction (two-body collision) and is essentially independent of temperature (λ adsorbs well on ice, at room temperature, or at 37°C). For titering λ, it is prudent, although not absolutely essential, to use maltose-grown cells. This is very important, however, when titering in vitro packaging extracts that contain a large excess of phage tails. For making liquid lysates, where many rounds of growth are required, less-efficient adsorption is useful, so use cells grown in the absence of maltose in TB medium.

Adsorption versus Injection

Bacteriophage λ adsorption requires the LamB protein and Mg^{++}. Ca^{++} is a competitive inhibitor. The cells can be dead (e.g., poisoned with cyanide) and λ will still adsorb. Phage λ will adsorb to debris if it contains the LamB protein. High salt (greater than 0.5 M) inhibits adsorption. Injection of λ requires energy and does not occur at room temperature. Consequently, λ will not form plaques at room temperature. Infections can be synchronized by adsorbing on ice or at room temperature and then diluting into warm media to raise the temperature.

Storage of λ

The most convenient and time-proven method for storing λ is by preparing a plate stock (see Procedure 2, p. 91). If kept in a tightly sealed tube or bottle at 4°C, plate stock lysates last for years. Lysates are sterilized with chloroform but should not be stored over excess chloroform. They should be stored in the dark; light may react with the lysate over time, creating photoproducts that kill λ. Lysates should be as free of bacterial debris as possible to reduce loss of titer through adsorption. If contamination appears, the lysate should be remade. The method of choice is to add a few drops of chloroform, shake at 37°C for 5 min, and then clear the lysate by low-speed centrifugation. Plaque purify the phage and prepare a fresh plate stock.

Chloroform is not entirely benign with respect to λ. Chloroform undergoes photolysis, yielding compounds lethal to λ. Fresh chloroform is best, although one should extract the chloroform with anhydrous sodium bicarbonate to remove undesired photoproducts before use. Chloroform should be kept in amber glass bottles away from direct light. Since λ will remain active in pure chloroform, be careful; a phage-contaminated chloroform bottle can wreak havoc in a genetics lab! The use of chloroform can be avoided by filtering phage lysates through a 0.45-μm membrane; however, this is expensive and may cause a drop in titer.

The presence of small amounts of agar in plate stocks apparently aids in lysate stability during storage. It is commonly observed that highly purified phage are quite unstable unless suspended in a protecting medium. A useful, defined suspension medium for highly purified phage is 0.01 M Tris·HCl, 0.01 M $MgSO_4$, 0.05 M NaCl, pH 8.0. An excellent, but undefined, suspension medium is TB medium containing a 1:4 dilution of molten top agar and 0.01 M $MgSO_4$.

Permanent storage of λ is best achieved by freezing at −70°C or in liquid nitrogen. The method of choice is as follows. Use screw-capped, plastic freezing vials (2-ml pro-vials or Nunc tubes) or 1-dram Wheaton glass vials (see Appendix D, p. 231). **Note:** Do not use glass vials in a liquid nitrogen freezer!

In one vial add:

70 µl DMSO
1 ml phage lysate

Mix. Label.

Freeze two vials per lysate. Store in two separate locations, if possible. In this media, λ can be frozen and thawed repeatedly with little or no loss of viability. It is best, however, to open a frozen vial and scrape off a little frozen lysate with a sterile loop or toothpick. Resuspend the lysate sample in a 50-µl droplet of λ-dil and streak or plate out for phage on an appropriate host. Return the storage vial to the freezer so that it is never allowed to completely thaw.

Since λ is a temperate phage, it can also be stored in a lysogen (see Appendix D). Finally, with the advent of in vitro packaging, λ can also be stored as DNA (see Procedure 39, p. 173). Phages can be recovered by packaging an aliquot of the desired DNA sample.

APPENDIX F
Phenotypes and Genotypes of λ

The genetic map of bacteriophage λ is highly developed (Hershey 1971; Echols and Murialdo 1978; Szybalski and Szybalski 1979; Hendrix et al. 1983). There are more than 50 specific genes or sites. Some of the terminology can be confusing, so below we define several of the commonly used terms that historically have been sources of problems.

λ is a *temperate phage*; i.e., it has two life cycles. In one, it is a *lytic phage* and kills cells after infection, liberating a burst of new phage. In the other, it is a *lysogenic phage*, integrating into the E. coli chromosome, turning off its lytic functions, and existing as a quiescent stretch of DNA called a *prophage*. The cells containing the prophage are called *lysogens*. If the prophage is integrated into the chromosome, the lysogens are usually stable; the prophage is inherited as a normal E. coli gene. If, for some reason, the prophage is not integrated, the lysogen is usually unstable, losing the nonreplicating prophage after several cell divisions. The prophage is kept quiescent by binding of the λ *repressor* (the product of the cI gene) to the left (o_L) and right (o_R) λ operators. When the repressor is inactivated by a process called *induction*, the operators are uncovered, transcription by E. coli RNA polymerase begins, and the lytic cycle is started.

Lysogens can be recognized because they contain λ repressor, which is made continually by the prophage. Phage λ can infect lysogens but cannot grow because the repressor inhibits the replication of the incoming DNA. Lysogens are therefore *immune* to *superinfection* by the same phage as the prophage. There are closely related *lambdoid phages* (e.g., 434, φ80, φ82, 21, PA-2) that have different operators and repressors. They can be classified by their repressors, i.e., by immunity of the corresponding lysogens. Phages with the same repressors and operators are said to be *homoimmune* (they cannot grow on lysogens of each other). Phages with different repressors and operators are said to be *heteroimmune* (they can grow on lysogens of each other). Phage λ and 434 are heteroimmune—λ will grow on a 434 lysogen and vice versa. It is possible to replace the immunity region of λ with the immunity regions of other lambdoid phages by genetic recombination. An example is λ*imm*434. This λ hybrid is now heteroimmune with respect to λ and homoimmune with respect to phage 434.

Synthesis of phage λ repressor (the cI gene product) is subject to control by several genes, including itself. Two genes, cII and cIII, are required to *establish* a lysogen. cII protein activates transcription of the cI and *int* genes as well as represses transcription of certain lytic genes. cIII protein works with cII, probably to inhibit a host protease that normally destroys cII protein.

Phage λ normally makes turbid plaques. The cells growing in the center of a plaque, making it turbid, are almost all lysogens. There are *clear-plaque* mutants of λ. These mutations affect functions that are involved in the *maintenance* or *establishment* of the lysogenic state. cI, cII, and cIII mutants form clear plaques. cI mutations are typically *recessive*; i.e., if repressor is supplied in *trans* to λcI mutants (from a homoimmune phage), lysogeny can be maintained. cII and cIII mutations affect establishment of lysogeny but have little effect on maintenance. It is possible to construct stable cII or cIII lysogens by supplying the cII or cIII gene product in *trans* during the initial infection. Mutations in the *site of action* of λ repressor (the o_L and o_R operators) no longer bind repressor. These mutants are said to be *virulent* (λ*vir*); they can grow lytically in the presence of homoimmune repressor. λ*vir* also makes clear plaques because it cannot lysogenize.

A useful, clear mutation in λ is *cIts857*. This cI mutation creates a temperature-sensitive repressor. At low temperature (30°C), λ*cIts857* makes a turbid plaque; at high temperature (37–42°C), it makes a clear plaque. Lysogens can be made and maintained at 30°C and induced efficiently at 42°C.

When λ lysogenizes *E. coli*, it must repress itself and, to be stably inherited, it must integrate into the *E. coli* chromosome. The λ *int* gene makes a protein required for insertion of λ DNA into *E. coli*. λ*int* mutants can repress themselves but cannot integrate. They therefore make turbid plaques, but the bacteria in the plaques are unstable or *abortive* lysogens.

Phage λ is a parasite, and as such it requires a number of *E. coli* functions for growth. These functions are involved in adsorption, replication, transcription, integration-excision, and capsid formation. Not all *E. coli* strains can support λ growth. It is useful to distinguish *E. coli* strains that cannot adsorb (but could grow λ if it could enter the cell) from those that adsorb λ but cannot support λ development. *E. coli* strains that cannot adsorb λ are called λ *resistant*; strains that cannot support λ development have a variety of names (e.g., *gro* mutants). It is important to contrast the difference between a λ-resistant *E. coli* (no adsorption) and a lysogenic strain (λ adsorbs and injects DNA but is prevented from growing by repressor). These two cell types can be distinguished operationally by using λ*vir* and λcI. λ*vir*, but not λcI, grows on a λ lysogen. Neither λ*vir* nor λcI grow on λ-resistant *E. coli*.

Phage λ requires the product of the *lamB* gene (LamB protein) for adsorption; λ adsorbs to this outer membrane protein through the λ tail fibers (gene *J*). Several classes of *lamB* mutants that produce an altered LamB protein no longer adsorb λ. It is possible to mutate the λ *J* gene so that the new tail fiber interacts productively with the altered LamB protein. These λ mutants that grow on strains normally resistant to λ adsorption are called host-range mutants (λ*h* mutants).

Phage λ carries its own generalized (homologous) recombination system (*red*). There are two genes, *exo* and *bet*, in the *red* system. The mechanism of *red*-promoted recombination is distinctly different from the *E. coli recA* system (Gottesman and Yarmolinsky 1968). Phage λ also carries an inhibitor (the product of the *gam* gene) of exonuclease V, the *E. coli recBC* gene product. The *red* and *gam* genes are adjacent in the λ chromosome and play additional roles in λ replication. Their functions are only partially understood. λ*red* mutants are defective in recombination and can be distinguished from wild type by their inability to form plaques on certain *E. coli* polymerase I (e.g., *polA1*) mutants or ligase (*lig7ts*) mutants. λ*red gam* mutants cannot form plaques on *E. coli recA* strains. This phenotype is called Fec$^-$. Wild-type λ is Fec$^+$. λ*red gam* mutants do grow, however, on *E. coli* strains lysogenic for the non-related phage P2. Phage λ does not grow on a P2 lysogen. This phenotype is called Spi (sensitive to P2 inhibition); λ*red gam* are Spi$^-$, and λ is Spi$^+$.

The genes essential for viability of λ are designated by capital letters (*A*, *B*, etc.). Mutations that confer a clear-plaque phenotype are designated *c*. Host-range mutations are designated *h*. In general, other genes are given a three-letter notation (*int*), using a system similar to that used for *E. coli* (see Appendix B, p. 223). For a useful compilation of genes and products, see Szybalski and Szybalski (1979).

Many λ mutants are conditional. Such mutants require certain host strains or culture conditions for growth (*permissive* as opposed to *nonpermissive*). Two common phenotypes are temperature sensitivity and dependence on tRNA suppressors. The latter mutants contain chain-terminating, nonsense mutations and require hosts with nonsense suppressors to grow. The most frequent type of nonsense mutations in λ are amber mutations. Traditionally, in λ genotypes, amber mutations are designated am (e.g., *P*am).

APPENDIX G
Titering Phage Lysates

In experiments using bacteriophage, it is often necessary to quantitate the number of plaque-forming units (pfu) per milliter of lysate. Described below are two simple methods for determining phage titers. Both methods require serial dilution of a phage stock or a pickate.

A pickate is prepared by using toothpicks to stab single plaques to a fresh lawn of host bacteria. Following incubation at 37°C for 4–6 hr, the phage can be harvested by plugging the entire lysed area around the stab with a 1-ml disposable, plastic pipette and by suspending the agar plug in 1 ml of TB with no Mg^{++}. Add a drop of chloroform. This suspension should contain $10^8 - 10^9$ phage.

Serial Dilution Method

1. Prepare, in advance, culture tubes containing 5 ml of sterile λ-dil.

2. To 5 ml of λ-dil add 5 µl of lysate and mix. This is a 1/1000 dilution, or 10^{-3} in shorthand. Using a fresh pipette, take 5 µl of the 10^{-3} dilution and add it to 5 ml of λ-dil. This is a 1/1,000,000 dilution, or 10^{-6}. These two dilutions should be sufficient to titer most stocks.

Spot Titer Method

This is semiquantitative and useful for screening several lysates because many can be done on one plate.

1. Make a lawn of cells by adding 0.1 ml of a $MgSO_4$ cell suspension to a small, sterile tube, followed by 3 ml of molten (45°C) TB top agar. Immediately pour the mixture over a TB plate. Tilt the plate to ensure that the top agar mixture is spread evenly over the entire surface. The top agar will solidify in a few minutes.

2. After the top agar layer has solidified, spot one drop each of the 10^{-3} and 10^{-6} dilution on the lawn. A drop of about $5-10$ μl is sufficient, forming a spot no larger than the size of a dime. With practice and experience, many spots can be put on a plate. Blue-tipped, nonheparinized hematocrit capillaries are "sterile" and inexpensive and deliver about the right size drop upon touching the plate. An alternative that can be more accurate is to use one of the many automatic micropipetters available. Use caution, however, because the micropipetters can become contaminated and spread phages where you do not want them.

3. Incubate the plate at 37°C overnight.

4. The 10^{-3} spot should be completely lysed. For a clear phage, some phage-resistant colonies may appear in the lysed area. There should be single plaques in the 10^{-6} spot, if the lysate is good. Ideally, there should be 50–100 plaques in the 10^{-6} spot for a phage lysate with a titer of 10^{10}/ml.

Quantitative Titer Method (Duplicate Plates)

1. Set up six small test tubes in a 30°C temperature block for each titration.

2. Add 0.1 ml of a $MgSO_4$ suspension of cells to each tube.

3. From the 10^{-6} dilution first, add 5 μl to two tubes and 50 μl to two more. From the 10^{-3} dilution, add 5 μl to two tubes.

4. After 5 min, add 3 ml of molten (45°C) TB top agar to each tube.

5. Immediately pour the mixture over a prelabeled TB plate.

6. When the top agar has solidified, incubate at 37°C overnight.

7. Count the plaques and calculate titer. For example, from the 10^{-6} dilution:

$$\frac{\text{No. of plaques}}{\text{Vol. plated (ml)}} \times 10^6 = \text{pfu/ml in lysate}$$

Note

On *spot titer plates*, the plaque size may vary due to different rates of adsorption. The plaque size on *quantitative titer plates* should be much more uniform and typical of each individual phage isolate.

APPENDIX H
Spontaneous Induction and the Release of Phage from λ Lysogens

The temperate bacteriophage λ can stably lysogenize *E. coli* by several pathways. Despite the fact that the lysogens are formed by different mechanisms, all will spontaneously release phage. This can be best demonstrated by treating an overnight culture of a lysogen with chloroform, removing cellular debris by centrifugation, and showing the presence of viable phage in the supernatant. If one understands the process of spontaneous prophage induction, it becomes an important genetic tool.

There are four classes of λ lysogens described in this manual. First is a wild-type λ lysogen formed by site-specific recombination at the bacterial attachment site, *att*B, using the phage attachment site, *att*P. The second is a lysogen formed by homologous recombination. The homology for recombination is either bacterial DNA carried by the phage or phage DNA carried in the bacterial chromosome. The third lysogen described is a λ phage defective in excision that is integrated normally by site-specific recombination. Such a prophage is "locked in"; i.e., it cannot excise from the chromosome by site-specific or homologous recombination; an aberrant event is required (e.g., a λ*int* or a λ*xis* mutant phage that has been integrated by using a helper phage [see Procedure 35, p. 166]). Fourth, a λ phage integrated by a transposition event is described. Such phage are also "locked in"; they cannot excise. An example is the λ phage that lies adjacent to a gene fusion constructed as described in Experiment 1 (p. 7); the site of integration of this phage was originally determined by a Mu transposition event.

Each of these four classes of lysogens will release phage spontaneously at a characteristic level. Consider Table 4 on page 242.

One can understand the spontaneous release of phage from each of these lysogen classes by knowing that two independent events must occur. The number of phage released is determined by the product of the probability of each of these two events. First, the repressor must be inactivated. Second, the prophage must be excised from the chromosome. Both events must occur to release phage. Either one alone is not sufficient. The first event, inactivation of repressor, is considered a constant for all four lysogen classes. Repressor is inactivated spontaneously in about 1 out of every 10^5 cells (10^{-5}). Excision of the prophage varies in each class. In class I, once repressor is inactivated, excision occurs by the extremely

Table 4
Yield of Plaque-forming Phage in Supernatants of Uninduced Overnight Cultures

Lysogen		Phage/ml
Class I:	wild-type λ lysogen	10^6
Class II:	lysogen formed by homologous recombination	10^4
Class III:	"locked-in" lysogen (e.g., λint or λxis)	10
Class IV:	"locked-in" lysogen (e.g., Fig. 2)	10

efficient, site-specific λ recombination system. The efficiency is essentially 100%. In class II, excision occurs by the reverse of the integration event, which was by homologous recombination. The frequency of homologous recombination (i.e., excision) is dependent on the amount of homology involved. Generally, no more than 1 in 1000 (10^{-3}) to 1 in 100 (10^{-2}) of these lysogens will have an excision event due to homologous recombination. Class III and IV lysogens are "locked in." They cannot exise by any event other than some very rare aberration. Such atypical excisions occur no more than 1 in 10,000 cells (10^{-5}). In all cases, the yield of phage per cell in a normal λ replication cycle is about 100 (10^2).

To obtain the estimated phage yield per milliliter of uninduced overnight culture, multiply the last column (product) in Table 5 by the cell density of an overnight culture (typically 10^9/ml) and multiply that product by the yield of phage per cell (typically 10^2/cell).

It is important to remember that λ repressor inactivation requires the $recA$ gene product. In $recA$ cells, the frequency of repressor inactivation drops from 10^{-5} to less than 10^{-8}. Phage λ cannot be induced in a $recA$

Table 5

Lysogen class	Probability of repressor inactivation	Probability of excision	Product
I	10^{-5}	1	10^{-5}
II	10^{-5}	$10^{-2}-10^{-3}$	$10^{-7}-10^{-8}$
III	10^{-5}	10^{-5}	10^{-10}
IV	10^{-5}	10^{-5}	10^{-10}

strain.[1] However, if a temperature-sensitive repressor mutation is used (e.g., cIts857), then class I recA λcIts857 lysogens can be induced simply by heat. In addition, the recA gene product is also absolutely essential for homologous recombination. Thus, the excision frequency of class II recA lysogens drops to less than 10^{-8}. Therefore, the probability of the spontaneous release of phage from such a lysogen is 10^{-16}.

---

[1] In overnight cultures of class I recA lysogens, however, there are considerably more phage in the supernatant than as calculated above. These phage almost always make clear plaques and are from those rare lysogens in which the repressor gene itself was mutated. A typical number for spontaneous mutation frequency is 10^{-6}.

APPENDIX I
Cloning Vectors

With the advent of recombinant DNA technology, numerous vector systems have been developed. Choosing an appropriate cloning vector can be a difficult task. Each vector has its own characteristics that determine the most appropriate applications. Recent and useful summaries of many commonly used vectors can be found in Maniatis et al. (1982).

All cloning vectors have several features in common. These include (1) unique site(s) for insertion of heterologous DNA sequences, (2) the ability to replicate autonomously in a host cell, and (3) a replication cycle to facilitate rapid, physical separation of the vector DNA from the host chromosome. There are several broad categories of *E. coli* vectors. These include (1) phage vectors (Williams and Blattner 1980), (2) plasmid vectors (Bernard and Helinski 1980), (3) single-stranded phage vectors (Barnes 1980), and (4) cosmids (Hohn and Hinnen 1980).

IMPORTANT FEATURES OF THE VARIOUS CLONING VECTORS IN BACTERIAL GENETICS

Phage Vectors

Advantages. Most of the available phage vectors are derivatives of the bacteriophage λ. This phage has been studied extensively, and consequently numerous "tricks" have been developed to facilitate the cloning and identification of heterologous DNA sequences. These include vectors that rely on insertion of DNA fragments to form viable phages (e.g., Charon 4A; Blattner et al. 1977) as well as vectors that exhibit insertional inactivation of a screenable function (e.g., λD69; see below).

The technique of in vitro packaging provides a distinct advantage for λ vectors (see Procedure 39, p. 173). Since infection is much more efficient than transfection (uptake of λ DNA), construction of complete genomic libraries is simplified. Moreover, since it is straightforward to screen plaques for a particular DNA sequence by hybridization (see Experiment 6, p. 43) or for a particular gene product by using specific antibodies (Skalka and Shapiro 1976), it is not difficult to probe an entire genomic library by using a minimum of materials.

Perhaps the most important advantage of λ vectors is their application in genetic analysis. Phage λ vectors simplify the construction of plaque-forming transducing phages (see Experiment 5, p. 39). In addition, simple techniques have been developed for the isolation of deletions and mutations in cloned fragments (see Experiments 11, 12, pp. 71, 75). The existence of vectors with or without the phage site-specific integration system or vectors in which this system can be activated when required (e.g., λD69) expands the usefulness of recombinant DNA in classical genetic analysis. In particular, single copies of genes can be introduced in *cis* and *trans* in natural and unnatural locations with ease.

The λ life cycle itself can offer a useful, but not always obvious, advantage. When λ infects *E. coli*, the lytic cycle is finished in less than an hour. The DNA cloned in a λ vector is exposed to replication and selection in discrete time intervals that can be controlled by the investigator. Moreover, rare events are "saved" in phage particles that can be analyzed in detail. Single-burst experiments, one-step growth curves, analysis of defective transducing phages, in vitro packaging, and CsCl density gradients are only a few of the many tools available for the phage biologist to analyze the fate of foreign DNA carried by λ cloning vectors.

Finally, and of considerable value, the complete DNA sequence of wild-type λ is known (Sanger et al. 1982).

Disadvantages. Phage vectors are relatively large molecules. The minimum number of genes required to obtain a plaque-forming phage as well as the constraints on packaging the genome into infective particles dictates a minimum vector size of about 38 kb. Therefore, vector DNA always represents a high percentage of the DNA in any clone. Finally, because the usual "selectable" phenotype is the phage plaque itself, any insert that interferes with plaque formation will likely be underrepresented in a library or missed altogether.

Plasmid Vectors

Advantages. Since the discovery of plasmids and the fact that many encode resistances to toxic compounds, there has been a tremendous research effort in plasmid molecular biology. One outcome of this intense development was the adaptation of various plasmids as cloning vectors. In general, plasmid vectors bestow easily selectable phenotypes to any host cell. The minimal requirement for plasmid maintenance involves a small region of the plasmid, the origin of replication. This, combined with a selectable marker, is all that is required for a plasmid vector. As a result, many of these vectors are small (about 10% of the size of λ phage vectors). Thus, the vector often represents a small percentage of the DNA in a plasmid clone. Accordingly, it is easy to obtain large amounts of insert DNA for physical analysis. In addition, since there are fewer restriction enzyme cleavage sites in plasmid vectors, genetic engineering is simplified. The DNA sequence of many plasmid vectors is available.

Disadvantages. Unlike the discrete life cycle of phage particles, plasmids must be maintained as constantly replicating molecules in a host cell. This represents a continuous source of selection. The DNA, once isolated, can be stored for long periods of time, but this requires special treatment not necessary for phage vectors.

Many plasmid vectors exist in the cell at a high copy number. This can lead to amplification of cloned gene products. Although often useful, such expression, if detrimental to cell viability, can lead to either alterations in a particular clone (mutations) or selection against that clone (underrepresentation in a library). Moreover this amplification is nonphysiological and can complicate genetic analysis.

Plasmids can be integrated into the chromosome only under unusual conditions. This requires the use of particular host-cell mutants (*polA*; Greener and Hill 1980). It is often difficult to take advantage of this property, which limits applications requiring homologous recombination with the host chromosome.

Single-stranded DNA Phages

Advantages. Inserts in single-stranded DNA phage vectors are packaged as single-stranded DNA in particles. Consequently, these vectors find their widest application in DNA sequence analysis, in vitro mutagenesis, and for construction of pure, single-stranded probes. Such phage vectors grow to remarkably high titers, making the isolation of large amounts of DNA a very simple task. The entire DNA sequence of the most commonly used phage, M13, is known (van Wezenbeek et al. 1980).

Recently, plasmids carrying the origin of replication from single-stranded DNA phages have been constructed. These hybrid replicons behave as plasmids but have a further advantage in that they can be used as a convenient source for single-stranded DNA (Dente et al. 1983; Zagursky and Berman 1984).

Disadvantages. Single-stranded DNA phage vectors replicate as plasmids in the cytoplasm of infected cells. Since they do not kill the cell by lysis but are assembled and secreted through the host membranes, they obviously affect host-cell physiology. In most cases it is prudent to avoid analyzing gene expression from inserts in such vectors.

Cosmids

Advantages. Cosmids are unusual hybrids of phages and plasmids. As the name implies, cosmids carry the *cos* site of phage λ, a plasmid origin of replication, and a selective marker(s). They are small and can accommodate large DNA inserts (up to 50 kb). Cosmids replicate as plasmids but because they have a *cos* site, they can be packaged into phage particles either by in vitro packaging of purified cosmid DNA or upon infection of cosmid-containing cells by a λ helper. Cosmids have proven applications to DNA cloning experiments in which long, contiguous stretches of DNA must be cloned.

Disadvantages. Although these plasmids can be packaged into phage particles, they are basically very defective transducing phages. They do not form plaques and lack every useful phage system except the *cos* site. Also, the required size of inserts defined by packaging constraints makes these plasmids very large and often unstable. Most genetic analysis is better done with λ and plasmid vectors.

VECTORS USED IN CONJUNCTION WITH THIS MANUAL

λD69

This vector was constructed by Mizusawa and Ward (1982). Its most useful features are (1) the presence of the λ site-specific recombination system, (2) a single cloning site for *Bam*HI fragments within a gene whose function can be visually screened, and (3) the immunity of phage 21 so that the vector can be used in conjunction with the heteroimmune λ fusion phages described in this manual.

λD69 is deleted for most of the nonessential DNA present in λ. Specifically, 10.4 kb have been deleted (a 21% deletion of λ). This means λD69 is almost as small as λ can be and still make a plaque. The map of λD69 is shown in Figure 22. The vector is fully proficient at site-specific and general recombination. It has a wild-type *imm*21 repressor (turbid plaques at all temperatures).

There are two particularly useful cloning sites in λD69. One is the single *Bam*HI site within the *int* gene and the other is the single *Hin*dIII site 300 bp to the left of *att*P. When the *Bam*HI site in *int* is used, recombinants can be visually screened by their Int⁻ white-plaque phenotype (see Procedure 5, p. 97; Fig. 15, p. 80). Such hybrids remain *att*P⁺ and can be made to integrate efficiently by supplying Int via a helper phage (see Procedure 35, p. 166). Lysogens of λD69 hybrids made in this way are "locked in"—they cannot excise because they are *int*.

When the *Hin*dIII site just to the left of *att*P is used, there is no visual screen for recombinant phages. The density selection is a useful procedure (see Procedure 33, p. 160) in this case. These λD69 hybrids are fully integration/excision proficient. The red-plaque test (see Procedure 5), can be used to screen deletion derivatives (see Procedure 19, p. 123) because deletions removing *int* function are very likely to enter the fragment cloned at the *Hin*dIII site.

The cloning capacity for either *Bam*HI or *Hin*dIII is fragments from 0 kb to about 13.9 kb. This is calculated as follows: λD69 is about 39.1 kb; the maximum size for λ is about 53 kb; therefore you could add 13.9 kb (53 kb−39.1 kb) to λD69 and not exceed the maximum size.

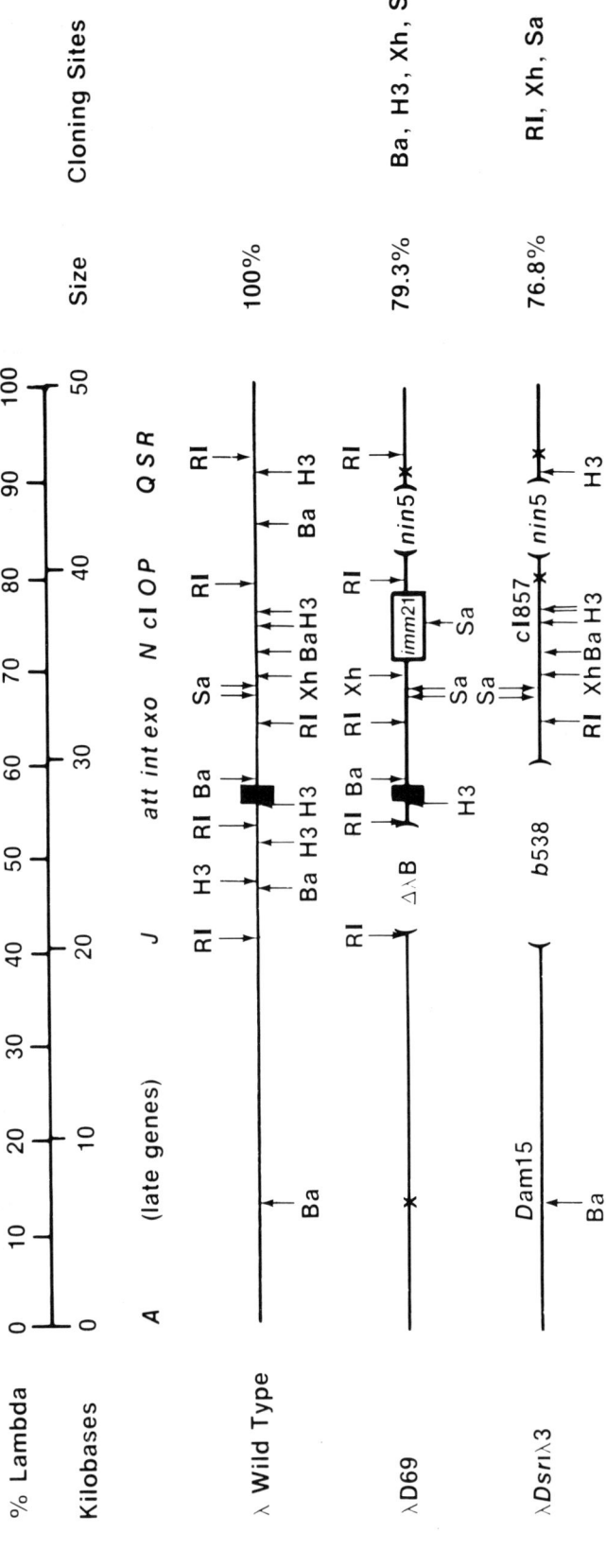

Figure 22
Structure of the λ cloning vectors described in this manual, λD69 and λDsrIλ3, along with a map of wild-type λ. Some λ genes are indicated. (■) att site. (RI) EcoRI; (Ba) BamHI; (H3) HindIII; (Sa) SalI; (Xh) XhoI. (✗) Mutation that destroys a restriction enzyme cleavage site. Parentheses indicate the extent of deletions.

λDamsrIλ3

This vector was constructed by Sternberg et al. (1977) primarily to exploit the unusual properties of D amber (Dam) mutations in λ. The most useful cloning site is the single EcoRI site in the exo gene. The XhoI and SalI sites can also be used. Most of the nonessential DNA has been removed (22.6%, or 11.2 kb), but in doing so the entire site-specific recombination system has been deleted. The map of λDamsrIλ3 is shown in Figure 22.

By inserting fragments at the EcoRI site in the exo gene, one can make use of several phenotypes of exo phages (see Procedure 36, p. 167). λ exo phages do not form plaques on E. coli strains carrying either the lig7ts or a polA mutation. This property can be used to distinguish hybrid phages from the parent. In conjunction, the density selection (see Procedure 33, p. 160) is valuable for use with this vector.

The Dam15 mutation is useful for the isolation of deletions and scoring for λ hybrids with inserts of less than 1.9 kb (see Procedure 20, p. 127). Phage λ will grow normally in the absence of the essential capsid protein gpD only if its DNA content is no greater than 82% (40.6 kb) of λ wild type. The vector λDamsrIλ3 contains less than this amount of DNA. Consequently, the vector grows equally well in strains with or without amber suppressors. Hybrid phages containing EcoRI fragments larger than 1.9 kb regain the requirement for gpD and now plate as a typical amber mutant; i.e., they only form plaques on strains with amber suppressors. This property can be exploited for isolation of deletions in cloned fragments larger than 1.9 kb simply by plating on Su$^-$ E. coli strains.

The cloning capacity of this vector is about 14.7 kb. This calculation is based on the size of λDamsrIλ3 of 38.3 kb and the maximum size for λ of about 53 kb.

pBR322

pBR322 (Bolivar et al. 1977) is the prototype of plasmid vectors. It is small (4363 bp) and the complete DNA sequence is known (Sutcliffe 1979; Peden 1983). It contains an origin of replication for ColE1 and genes that specify resistance to two antibiotics. One gene specifies a β-lactamase from Tn3 that bestows resistance to ampicillin and related compounds. Another region of pBR322 specifies resistance to tetracycline. (Neither of these genes is contained in a functional transposon.) There are many unique restriction enzyme sites allowing insertional inactivation of either resistance marker. The origin of replication is efficient and produces 20–50 copies of the plasmid per cell. This facilitates rapid preparation of DNA in amounts adequate for most experiments and often yields significant amplification of cloned gene products. Although we do not employ pBR322 in this manual, all of the plasmids used are derived from this vector. Indeed, this cloning vector is one of the most widely used of the recombinant vectors.

pMLB1034

A number of vectors have been developed for constructing protein fusions. These vectors contain the *lacZ* gene lacking a promoter, ribosome-binding site, and ATG initiation codon (Casadaban et al. 1980; Guarente et al. 1980; see Experiment 2, p. 18). Since the *lacZ* gene lacks its expression signals, β-galactosidase is not produced. However, when the 5' end of a gene containing expression signals and encoding the aminoterminal portion of a protein is joined to the *lacZ* sequence in the proper reading frame, a hybrid gene is formed that confers LacZ⁺. Thus, these vectors function by insertional activation of *lacZ*, which can be detected on indicator plates.

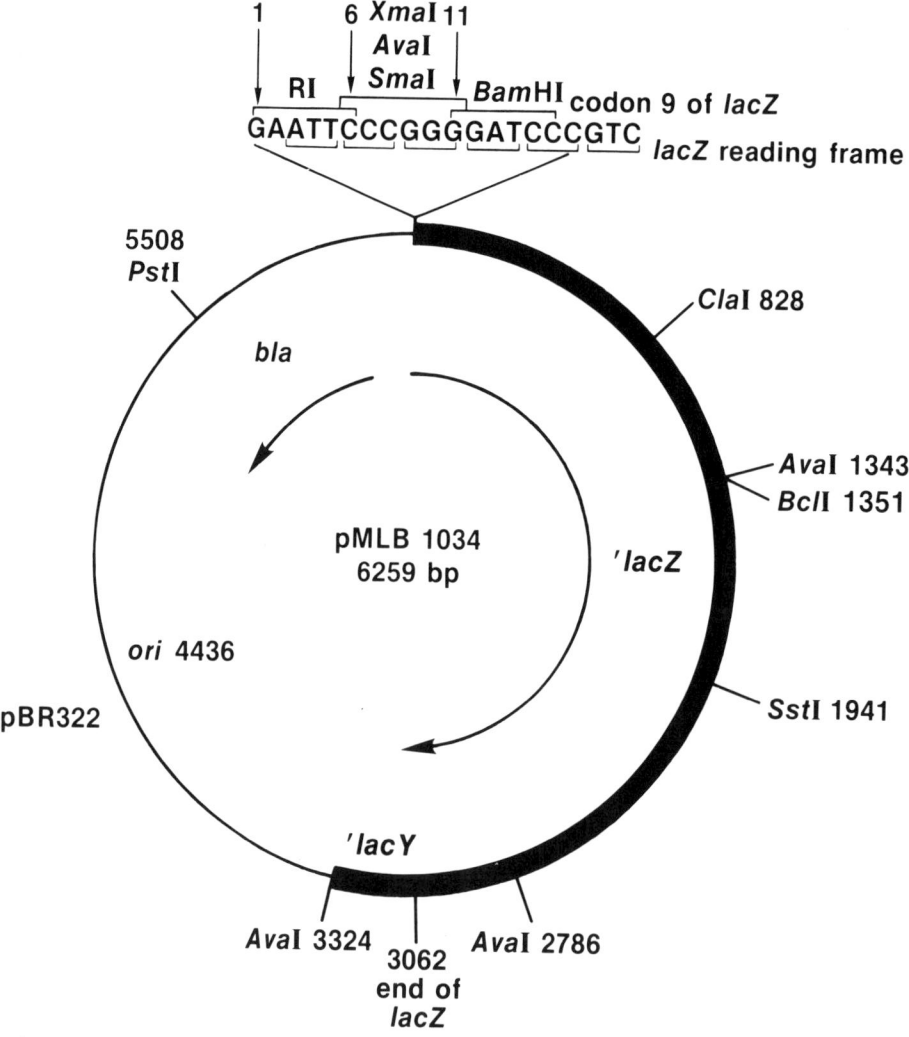

Figure 23

Structure of pMLB1034. (→) Direction of transcription of the β-lactamase gene (*bla*) and '*lacZ*. (■) DNA from the *lac* operon; (———) sequences from pBR322.

The vector described in this manual for the in vitro construction of gene fusions is pMLB1034 (see Experiment 2). The *lac* region of this plasmid is derived from pMC871 (Casadaban et al. 1980), but the sequences distal to the *Ava*I site in *LacY* (see Appendix M, p. 266) have been deleted. A map of this vector, including important restriction enzyme cleavage sites, is shown in Figure 23.

Two Vectors Designed for Cloning Gene Fusions

pMLB524. The techniques described in Experiment 1 (p. 7) and Appendix L (p. 261) for constructing *lac* fusions in vivo allow the isolation of specialized λ transducing phages carrying the fusion. The fusion can be further subcloned into a number of plasmid vectors specifically designed for this purpose. The map of restriction enzyme cleavage sites in the *lac* operon shows a number of sites that occur only once in *lacZ* (see Appendix M, Fig. 30, p. 282). Three of these sites have been used to construct derivatives of pMLB1034 that carry various 3'-terminal fragments of *lacZ*. Such vectors can be used to subclone fragments from the *lac* fusion transducing phages and reform the *lacZ* fusion on the plasmid.

The use of pMLB524 is described in Experiment 3 (p. 28). Plasmid pMLB524 (3251 bp in size) was constructed by *Eco*RI digestion of a pMLB1034 derivative in which the naturally occurring *Eco*RI site in *lacZ* is present. Only the terminal 53 bp from *lacZ* remain in pMLB524. This vector has been used to clone *Eco*RI fragments from various *lac* fusion transducing phages (Berman et al. 1984). Restoration of the *lacZ* gene is detected by using the indicator dye Xgal.

Two other cloning vectors have been constructed by deletion of pMLB1034 from the *Eco*RI site to the *Sst*I or *Cla*I site and insertion of small oligonucleotide linker DNA segments. The plasmid pMLB1060 (4348 bp) is deleted for the *lacZ* sequence up to the *Sst*I site (see Fig. 30), and plasmid pMLB1094 (5457 bp) is deleted up to the *Cla*I site. (For nucleotide sequence, see Appendix M, p. 266). These vectors are used in a manner analogous to pMLB524, to clone fusions by insertional activation of *lacZ*.

λ*NF1955.* The use of fusions to measure gene expression or to isolate mutants is best performed with a single copy of the fusion in the bacterial chromosome. Fusions constructed in multicopy plasmids often are not suitable for genetic analysis because of the high gene dosage (see Experiment 2, p. 18). To facilitate insertion of fusions into the *E. coli* chromosome, a vector, λNF1955, has been constructed (Hui et al. 1982; Fig. 24). This phage has a unique *Eco*RI site that can be used to clone fusions by insertional activation of *lacZ*. The portion of the chromosome to the left of this site is derived from λp1(209) (Casadaban 1976a) and carries the 3' end of *lacZ* (as described for pMLB524 above; see Appendix M) and a complete *lacY* gene. The portion to the right of the *Eco*RI site is from λgtWES·λC*, a derivative of λgtWES·λC (Enquist et al. 1976) in which the

EcoRI site in *exo* has been altered by mutation. Fusions can be cloned at the EcoRI site to produce an *att*⁺ transducing phage. The Sam mutation in λNF1955 is allele 100. Unlike Sam7, Sam100 is suppressed by both *supD* and *supF*. Sam7 is only suppressed by *supF*.

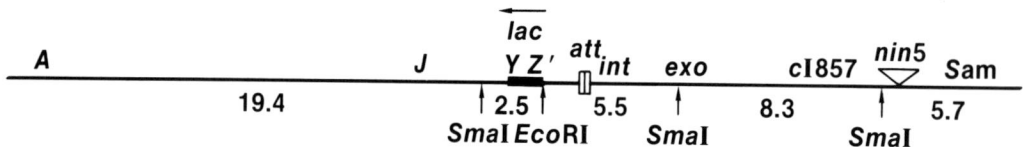

Figure 24
Structure of the gene fusion cloning vector λNF1955. The numbers refer to lengths in kilobases of indicated fragments. (▯) *att* site; (▬) the fragment from the *lac* operon.

APPENDIX J

Moving Mutations from One Replicon to Another by Recombination

A major goal of this manual is to demonstrate the power and utility of the combined approaches of gene fusions, recombinant DNA methods, and more classical genetic techniques. The successful combination of these methodologies requires the skillful use of recombination to move mutations from one replicon to another at will. For example, mutations that are isolated on specialized transducing phages (see Experiments 11, 12, pp. 71, 75) or plasmids (see Experiment 2, p. 18) must be transferred to the chromosome in order to assess reliably the mutant phenotype. Conversely, physical or DNA sequence analysis of mutations isolated on the chromosome is greatly simplified by first moving the mutation to a specialized transducing phage or to a multicopy plasmid vector (see Experiments 8, 9, 10, pp. 53, 59, 63). In addition, methods for recombining mutant alleles between replicons provide the tools essential for fine-structure genetic mapping and proper complementation analysis (see Experiment 13, p. 79).

What follows is a description of several methods for recombining mutant alleles from one replicon to another. These methods rely on genetic homology (homologous recombination). Some of these have been used in different experiments so the principles depicted refer to those specific examples. Although it is possible to embellish this catalog by incorporating in vitro recombination methods, the discussion will be limited to the most common in vivo situations.

Moving Mutations between a Specialized Transducing Phage and the Chromosome

The fundamental requirement for moving a mutation from one replicon to another is that the mutation must be located within a region of homology. The entire exchange involves a double, reciprocal recombinational event that can be viewed as the outcome of the two separate recombinational events, each of which has its own particular consequences.

The first reciprocal recombinational event results in the formation of a stable phage lysogen. Selections for these lysogens can be based on the expression of prophage immunity (see Procedure 6, p. 99) or on other dominant markers carried by the phage (e.g., a Tn9 in λpSG1 in Experiment 1, p. 7; Km^r determinant in λp1081.1 in Experiment 10). Provided that the phage integration is not mediated by a site-specific recombination system, lysogen formation will occur by the type of crossover depicted in Figure 25. The result is a strain that is diploid for a particular region of the chromosome. These tandemly repeated regions are separated by the prophage genome. The frequency of this recombinational event will depend upon the extent of homology between the phage and chromosome.

The second reciprocal recombinational event results in the excision of the prophage from the chromosome. This crossover can occur anywhere within the tandem segments of homologous DNA. (The role of homologous recombination in the liberation of phage from this type of lysogen is discussed in detail in Appendix H, p. 241). Depending upon the point of crossover, the excised phage will be identical with the original phage or, as depicted in Figure 25, it will now carry the allele from the chromosome.

The approach described in Figure 25 can be employed to transfer alleles from the chromosome to a phage or vice versa. Transducing phage carrying the chromosomal allele can be found among those spontaneously released from the lysogen by employing a suitable selection or screen.

If the goal is to transfer an allele from a transducing phage to the chromosome, strains that have lost the prophage must be isolated. Often, selections or screens for such strains are based on markers that should segregate with the prophage. For example, a high-temperature selection can be employed to isolate clones that have lost a λcIts857 prophage. Subsequently, an appropriate screen can be employed to identify recombinants carrying the desired allele. Alternatively, it may be possible to select or screen directly for the desired recombinant. For example, a screen can be employed to identify clones that express a recessive mutant phenotype. In most cases, these clones are the result of a recombinational event that removed the prophage.

In the transfer of markers from the chromosome to λ, it is not always necessary to isolate the intermediate lysogen. The frequency at which a transducing phage acquires a chromosomal allele is the product of the probability of the two required crossover events. If the extent of homology is large, this frequency can be as high as 1%. Thus, simply by growing a transducing phage on a mutant host and then employing a suitable selection or screen, the desired phage recombinant can be identified directly.

Moving Mutations between an F' and the Chromsome and Vice Versa—Homogenotization of a Recessive Allele

To understand the events involved in moving a mutation from the chromosome to an F', we need to discuss homogenotization in *E. coli*. According to one model, homogenotization involves recombination and segregation of genetic markers.[1] A double, reciprocal recombinational event will move a mutant allele from the chromosome to an F' while at the same time replacing the mutant chromosomal allele with the wild type. Because there are at least two copies of the chromosome and two copies of the F' present in the cell at the time of division, at some frequency the episome carrying the mutant allele will segregate with a chromosome carrying the mutant allele, resulting in a daughter cell that is a homogenote expressing the recessive mutant phenotype. This pathway of homogenotization can be viewed as the result of two separate crossover events. The first crossover event leads to integration of the F' and formation of an Hfr. The second leads to release of the episome from the chromosome, thus restoring the orginal, autonomously replicating F'. If the two crossovers occur on opposite sides of the mutant allele, the mutations will be transferred to the episome. Diagrammatically, this is analogous to phage integration and excision by homologous recombination (Fig. 25). The frequency of homogenotization is the product of the probability of the two crossover events. In *E. coli*, it can reach 10^{-4} per cell per generation if the regions of homology are substantial. Accordingly, homogenotes can often be recognized by screening a population for merodiploids that express the recessive mutant phenotype (Miller 1972).

Moving Mutations between High-copy-number Plasmids and the Chromosome or Vice Versa

Moving mutations from small, high-copy-number plasmids to the chromosome or vice versa is not straightforward. The major problem with this recombination stems from the relaxed ColE1 replicon present in pBR322 and its derivatives. This replicon cannot be stably integrated into the chromosome without affecting cellular viability. (This may be harmful in the same way that an unrestrained λ origin of replication is harmful in the chromosome [Experiment 10, p. 63].) This problem can be overcome using certain *E. coli* mutants (e.g., *polA1*; Greener and Hill 1980). In these strains, the ColE1 replicon cannot function. Accordingly, chromosomal integration is required for plasmid maintenance. To use this procedure routinely requires a considerable amount of strain construction. Consequently, we employ a tranducing phage intermediate. As discussed below, moving mutant alleles from plasmids to phages or vice versa is quite simple, and as you have already seen, moving mutations from the chromosome to a phage is no problem either.

Moving Mutations between High-copy-number Plasmids and a Specialized Transducing Phage or Vice Versa

Once again, the same set of recombinational events discussed in the previous sections are relevant here. For recombination between plasmids

---

[1] Homogenotization may also occur by a gene conversion. This mechanism is nonreciprocal.

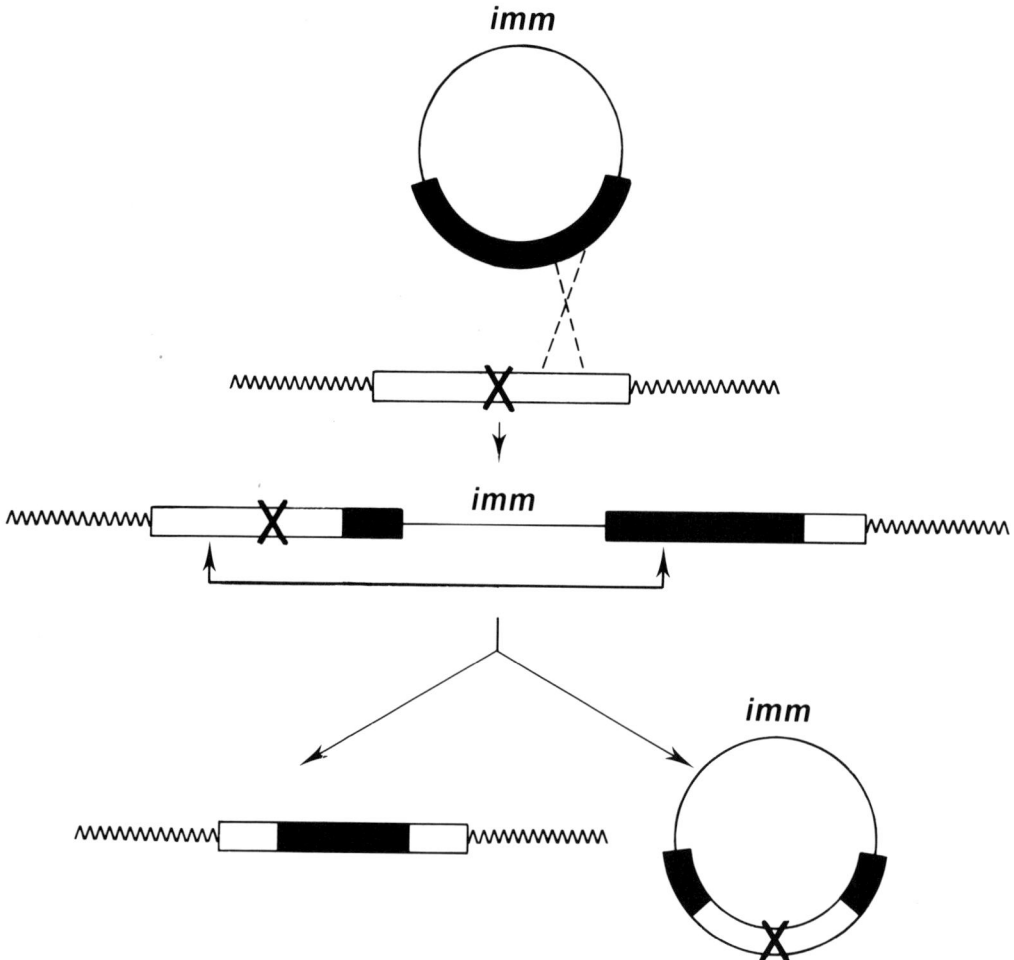

Figure 25
Moving mutations from the chromosome to a transducing phage. (———) Phage DNA; (ᗢᗢᗢ) chromosome. The homologous regions shared by the transducing phage and chromosome are shown as solid and outlined boxes, respectively. (**X**) Mutation.

and transducing phages, the scheme is outlined in Figure 26. A single, reciprocal recombinational event between a λ phage and a high-copy-number plasmid will result in the formation of a hybrid λ-plasmid structure. These structures are relatively stable and behave as large plasmids. If the size of this hybrid structure does not exceed the packaging limits of λ, it will be packaged into phage particles during lytic growth. Consequently, this lysate can be used to transfer the hybrid molecule to a suitable recipient strain by performing a transduction and selecting for a dominant marker carried by the plasmid (usually antibiotic resistance). In these transductants, a single recombinational event will resolve the λ-plasmid hybrid into its composite replicons (Fig. 26).

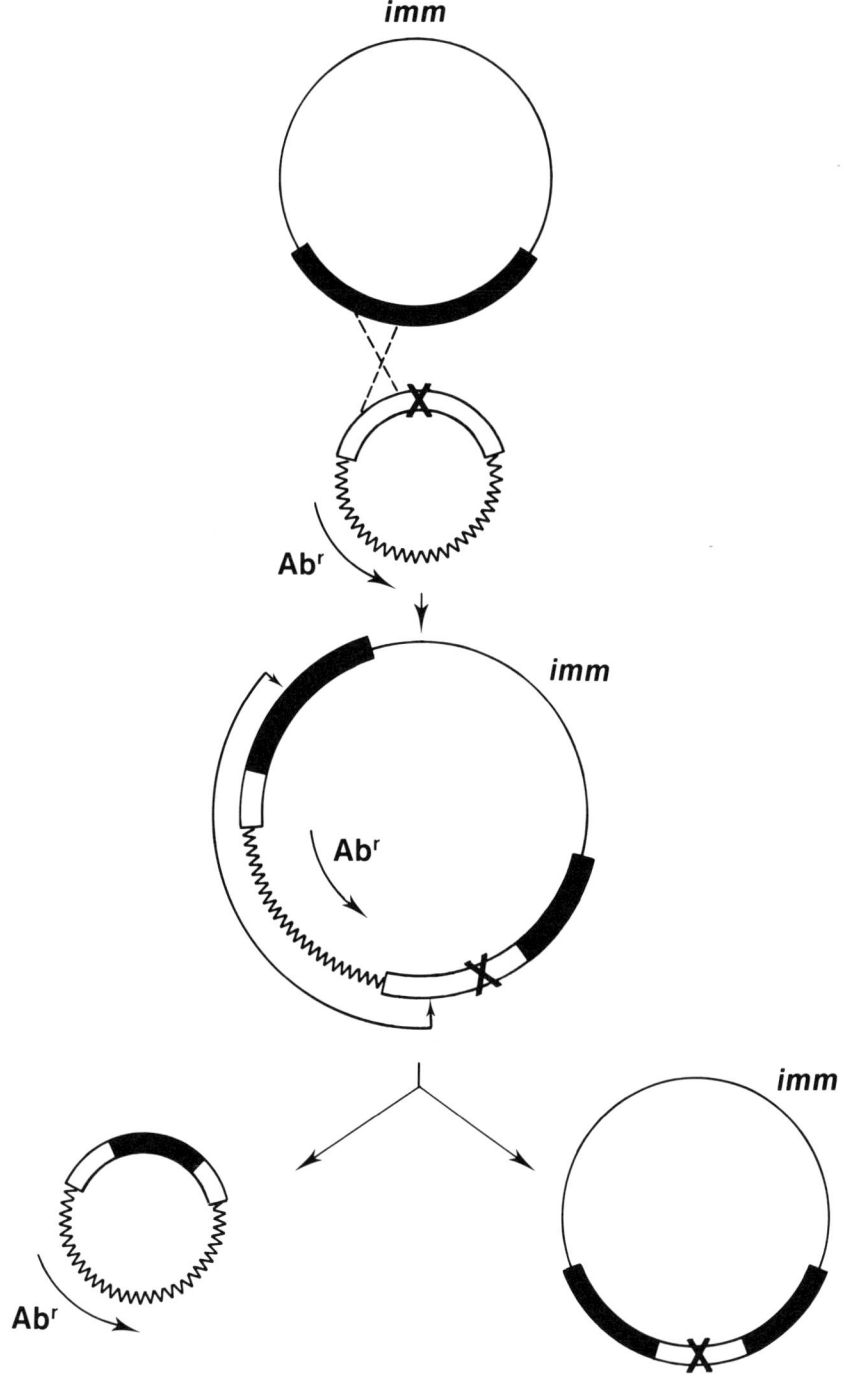

Figure 26
Moving mutations from a plasmid to a transducing phage. Symbols are defined in the legend to Fig. 25, except the jagged line here represents plasmid DNA.

Transducing phage, released from the transductants that have undergone the desired recombinational event, can be identified using an appropriate selection or screen. Such phage will not transduce markers associated with the plasmid. Plasmids can be recovered by preparing DNA from the transductants and transforming a suitable recipient. The intended recombinant can then be identified using a selection or screen. These transformants must be scored for the presence of λ genes (such as immunity; see Procedure 6, p. 99) to verify loss of the phage.

It is not necessary to package the λ-plasmid hybrid in order to transfer a particular allele from a plasmid onto a phage. Indeed, with large plasmids, it is not possible. Provided there is sufficient homology, transducing phage that have acquired the desired allele can be isolated directly, using an appropriate selection or screen, from a lysate prepared on the plasmid-containing host.

APPENDIX K
Genetic Verification of Gene Fusions

In general, the use of Mu*d*(*lac*, Ap) or λ*plac*Mu yields fusions of the *lac* genes to the desired gene; however, this is not always the case. Accordingly, it is essential to verify that expression of the *lac* genes in the fusion strain is dependent upon target gene sequences. If the regulatory properties of the target gene are known, then fusions can be verified by demonstrating that the *lac* genes have acquired the proper regulation. When the target gene is unregulated or the regulatory properties are not known, it is possible to devise a genetic experiment to demonstrate that *lac* is fused to the target gene. The essence of this genetic proof is that mutations that interfere with the expression of the target gene should also interfere with the expression of the *lac* genes once incorporated on the fusions. λ transducing phage carrying the gene fusion can be isolated simply (see Experiment 1, p. 7), providing a method for performing this test (Berman and Beckwith 1979; Debarbouille and Swartz 1979; Hall and Silhavy 1979).

The technique described for verifying a gene fusion requires a fusion transducing phage as well as a set of polar mutations in the target gene (see Experiment 8, p. 53). As illustrated in Figure 27, a transducing phage carrying a fusion between X and *lacZ* is introduced into a strain carrying a polar mutation in X. (This is pictured as a transposon [Tn] in X. Any polar mutation, including nonsense mutations, or promoter mutation will also work.) Since the recipient strain is deleted for the *lac* genes, and since the λ transducing phage used has no attachment site, integration into the chromosome occurs by homologous recombination. Two types of lysogens are possible if the mutation in X is promoter proximal to the X-*lacZ* fusion joint. Lysogenization will occur by either cross 1 or 2. After selecting lysogens (see Procedure 6, p. 99), the Lac and X phenotypes are scored. The occurrence of a Lac$^-$, X$^+$ lysogen is possible only if the transducing phage carries an X-*lac* fusion.

In principle, this method is no different than the mapping protocols described in Experiments 12 and 13 (pp. 75, 79) and the recombinational events illustrated in Appendix J (p. 253). Not only can it be used to verify gene fusions, but it can also be used to map polar mutations with respect to the location of a fusion joint and the promoter.

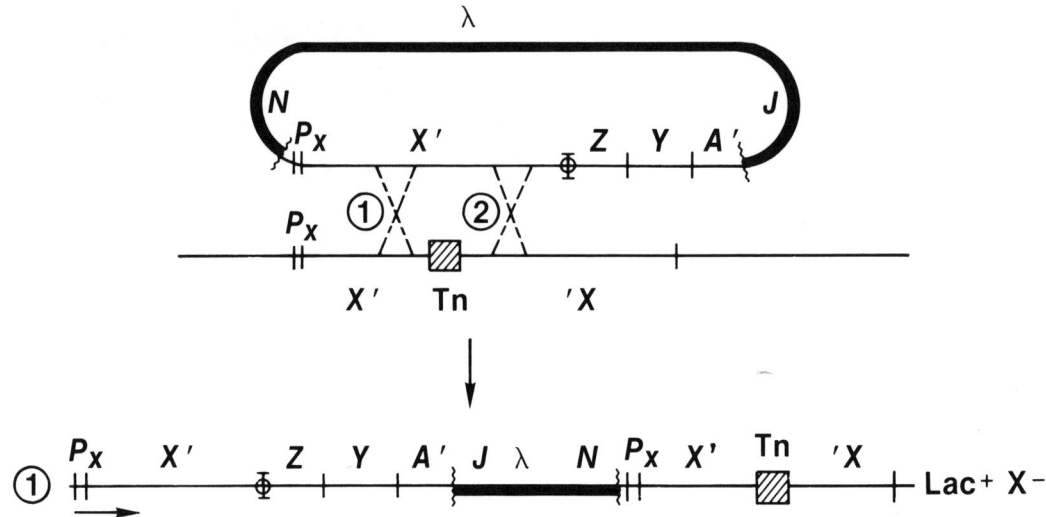

Figure 27
Verification of a gene fusion. Represented are the products of two different crossover events resulting in the chromosomal integration of a fusion transducing phage. (▬) λ DNA; (▨) transposon.

APPENDIX L
Using λplacMu1

A plaque-forming derivative of phage λ, λplacMu1 (Bremer et al. 1984), has been isolated that contains sequences from bacteriophage Mu enabling it to integrate into the E. coli chromosome by means of the Mu transposition system. The Mu DNA carried by this phage includes both attachment sites as well as the genes cI (specifies Mu repressor) and A (specifies Mu transposase). The phage also contains the lacZ gene, deleted for its transcription and translation initiation signals, and the lacY gene positioned next to the terminal 117 bp from the s end of Mu. Because of the physical structure of λplacMu1, the hybrid phage creates lacZ protein fusions in a single step when integrated into a target gene in the proper orientation and reading frame (Fig. 28). In contrast to strains carrying a Mud prophage, fusion strains constructed using λplacMu1 are stable and can be used for mutant selection and isolation of specialized transducing phages (see Experiment 1, p. 7).

The λplacMu1 phage was derived from a MudII(lac, Ap) insertion. After conversion of the Mu sequences to λ (see Experiment 1, Fig. 2, p. 12), λplacMu1 was formed by an aberrant excision event (Fig. 28). The capacity of this phage to transpose was recognized by its ability to transduce a Δ(lac) recA recipient to Lac$^+$.

We have chosen the maltose regulon as a model system to demonstrate the use of this phage (see Appendix O, p. 286).[1] A day-by-day protocol for isolating mal-lac fusions follows.

---

[1]In Experiment 1 (p. 7) we employed the ara regulon for gene fusion analysis. There are several technical reasons why the λplacMu1 phage cannot be simply employed for the isolation of ara-lac fusions.

262   Appendix L

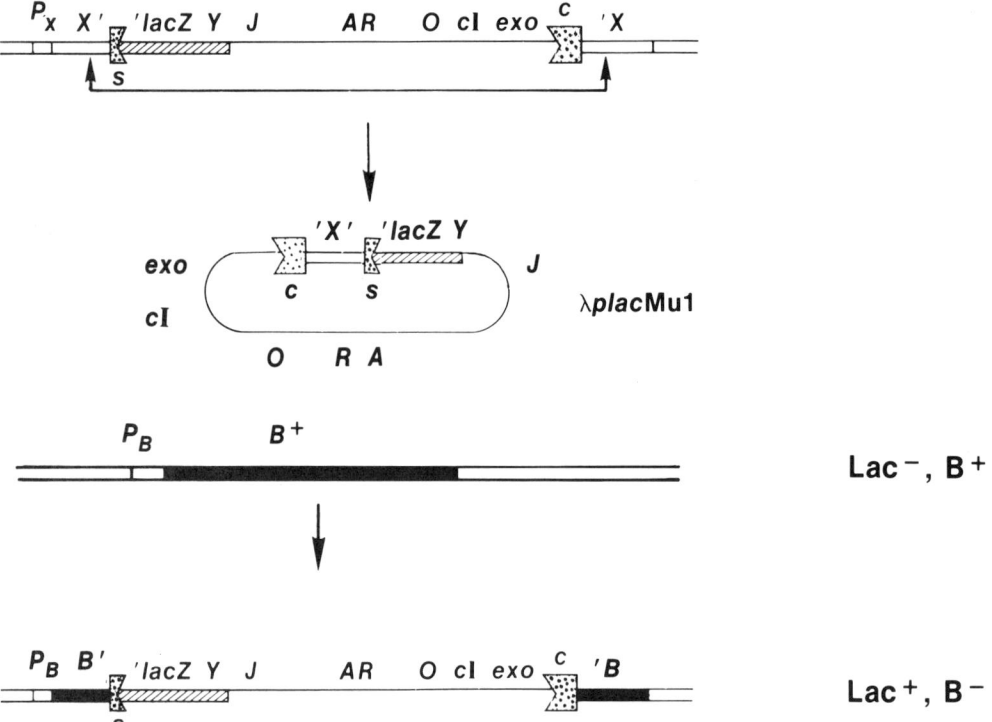

Figure 28
Origin and use of λplacMu1. The top line represents a Lac⁻ gene fusion constructed as described in Experiment 1 (Fig. 2, p. 12), using MudII (*lac*, Ap) inserted in the *ara* operon. λplacMu1 was formed by an aberrant excision event (bracket with arrows) that resulted in the formation of a transducing phage carrying both the c and s ends of Mu (▧), *lac* operon DNA (▨) without any expression signals, some flanking chromosomal DNA (▢), and all essential phage genes (———). If it inserts into the target gene (■) in the proper orientation and in the correct reading frame, it will generate a protein fusion that specifies a hybrid protein. Relevant phenotypes are listed at right.

DAY 1

Inoculate a 5-ml culture of MC4100 in L medium and grow overnight at 37°C.

DAY 2

Mix 1 ml of the MC4100 culture with 0.1 ml of λ*plac*Mu1 (10^8-10^9 pfu/ml) and λpMu507[2] (10^8-10^9 pfu/ml). Incubate for 30 min at 30°C. Add 5 ml of L medium and centrifuge (3500 rpm for 10 min) to remove unadsorbed phages. Discard the supernatant, resuspend the cells in 5 ml of L medium and centrifuge again. Resuspend cells in 1 ml of L medium. Prepare serial dilutions (10^{-1}, 10^{-2}, 10^{-3}, 10^{-4}, and 10^{-5}) in L medium. Plate 0.1 ml of the undiluted cells and each of the dilutions onto lactose minimal media. Incubate for 2 days at 37°C.

DAY 4

Replica plate Lac$^+$ colonies onto maltose tetrazolium agar. Incubate plates for 6–8 hr at 37°C.

Pick Mal$^-$ colonies (dark red) and purify on maltose MacConkey agar at 37°C. Use strain MC4100 as a control. Streak KLF41 on minimal glucose agar containing leucine, histidine, and methionine and grow at 37°C.

DAY 5

From the maltose MacConkey plates, pick Mal$^-$ colonies (white) and re-streak on maltose MacConkey, lactose MacConkey, and L agar plates containing Xgal at 37°C; include MC4100 as a control. Streak SE3001 and SV101 on L plates and incubate at 37°C overnight.

DAY 6

Score results. At this point it is prudent to prepare a P1*vir* lysate on the fusion strains and use this lysate to transduce a *lac* strain (MC4100) to Lac$^+$ by selecting for growth on lactose minimal agar (see Procedures 10, 14, pp. 107, 114). Transductants that acquire the fusion should also

---

[2] λ*plac*Mu is able to transpose by itself. However, transposition of this phage is stimulated at least 10-fold by coinfection with the helper phage λpMu507. The latter phage carries both the Mu A and B genes (Magazin et al. 1977). The increased transposition observed when using the helper phage probably results from the presence of the Mu B-gene product. To save plates and increase your chances of isolating a *mal-lac* fusion, use λpMu507.

acquire a Mal⁻ phenotype. By transducing the fusion to a clean genetic background, any problems associated with multiple λplacMu1 prophages can be avoided. To save time, these steps will be omitted.

The Mal⁻, Lac⁺ phenotype can arise from λplacMu1 insertions in either the *malA* or *malB* locus (see Appendix O). Insertions at these loci can be distinguished by a combination of phage sensitivity tests and complementation analysis. For the complementation analysis, we will use an F′ carrying *malA* but not *malB* and a λ transducing phage (λapmalB13) carrying the entire *malB* region. For the following tests, use blue colonies picked from the L agar containing Xgal.

1. Cross-streak Mal⁻ fusion strains against λvir and λcI at 37°C (see Procedure 6, p. 99).

2. Inoculate a 5-ml culture of KLF41 in L medium from the glucose minimal plate prepared on Day 4 and grow to mid-log phase. Centrifuge the KLF41 culture and resuspend in 2.5 ml of 10 mM $MgSO_4$. (**Caution:** Extensive vortexing may remove sex pili.) Introduce F′141 (*malA*⁺) into the fusion strains by cross-streaking the Lac⁺ fusion strains against strain KLF41 on maltose minimal agar at 37°C. Include MC4100, SE3001 (*malB*), and SV101 (*malA*) as controls. It takes 1–2 days to see a clear result.

3. Grow Lac⁺ fusion strains in 2 ml of glucose minimal media at 37°C to mid-log phase.

    Spot two 10-μl drops from each culture onto a maltose minimal plate. Let dry. Spot 10 μl of λapmalB13 onto one of these drops. Let dry and incubate at 30°C for 2 days.

    The results of these tests define the position of the Lac⁺ gene fusion. The various *mal-lac* fusion strains will have the phenotypes shown in Table 6.

Table 6

Gene(s)	λ**vir**	λ**cI**	Complementation to Mal⁺ by	
			F′141	**λapmalB13**
malK	r	r	−	+
malE,F,G	s	r	−	+
malP,Q	s	r	+	−
malT	r	r	+	−

(r) Resistance; (s) sensitivity; (+) growth; (−) no growth.

DAY 8

Unlike *ara-lac* fusions (see Experiment 1, p. 7), the induction of *mal-lac* fusions by maltose cannot be scored on indicator agar. Accordingly, β-galactosidase assays must be performed to verify the fusions. With strains that carry fusions in the *malA* region, this test can be performed directly; proceed to Day 10.

Fusions located in the *malB* region present a special problem. Since the fusion destroys a gene involved in maltose transport, maltose cannot enter the cell, and its addition will have no effect. To assay these fusions, use Mal$^+$ diploids. These should be picked from the Mal$^+$ colonies obtained by complementation with λap*malB13* (Day 6, step 3) and purified on maltose MacConkey agar.

DAY 9

Repurify Mal$^+$ diploids by streaking on maltose MacConkey agar. Be sure to check the Lac phenotype by streaking on lactose MacConkey agar. This is important because the fusion can be lost by homologous recombination with λap*malB13* (see Appendix J, p. 253).

DAY 10

Grow fusion strains (i.e., *malA-lac* or the Mal$^+$, Lac$^+$ *malB-lac* diploids) in glycerol minimal media with and without 0.2% maltose. Assay the β-galactosidase activity of the fusion strains according to Miller (1972). All *mal-lacZ* fusions should show at least a 10-fold induction of enzyme activity except *malT-lacZ*. The *malT* gene is not regulated in response to maltose addition. How might these fusions be verified?

λ transducing phages that carry the various *mal-lac* fusions can be isolated from the haploid fusion strains as described in Experiment 1, Day 11 (p. 17). Remember to add maltose instead of arabinose to induce the fusion. Techniques for cloning the fusions are described in Experiment 3 (p. 28).

APPENDIX M
The Lactose Operon[1]

The *lac* operon of *E. coli* is a cluster of genes at 8 min on the genetic map. The products of these genes are required for the cell to utilize the disaccharide lactose (glucose [β1−4] galactose). The production of the enzymes of the *lac* operon in a wild-type strain is induced by the presence of lactose in the medium. The gene order in the operon is *lacIPOZYA* (Ippen et al. 1968). Only the products of *lacZ* (β-galactosidase) and *lacY* (lactose permease) are required for growth on lactose (Lederberg 1947; Rickenberg et al. 1956; Cohn 1957; Fox and Kennedy 1965). The final gene in this operon, *lacA*, specifies a transacetylase (Zabin et al. 1959) that is involved in the detoxification of certain galactosides (Andrews and Lin 1976).

Transcription of the structural genes by RNA polymerase proceeds from the promoter site (*lacP*) and is negatively regulated by the Lac repressor, the product of the *lacI* gene (Pardee et al. 1959; Jacob and Monod 1961). Expression of the repressor is from a constitutive, low-level promoter. The Lac repressor binds at the operator site (*lacO*) to block transcription. Mutations in *lacO* result in constitutive production of the enzymes of the operon (Jacob et al. 1960). In addition to negative control, expression of the operon is dependent on levels of cAMP (Perlman and Pastan 1968), the product of the *cya* gene, adenylcyclase, and the cAMP receptor protein (CRP; Schwartz and Beckwith 1970). This positive regulation acts at *lacP*. Mutants able to express the structural genes in the absence of CRP or cAMP map in the *lac* promoter (Silverstone et al. 1970). A summary of the gene products and regulatory sites is shown in Table 7. For a collection of reviews of various aspects of the *lac* operon, see Beckwith and Zipser (1970) and Miller and Reznikoff (1978).

ASSAY OF *lac* GENE PRODUCTS

The most common assay for these gene products uses the chromogenic substrate *o*-nitrophenyl-β-D-galactoside (ONPG). This compound is colorless in aqueous solutions. Hydrolysis, mediated by β-galactosidase, pro-

---

[1]Much of this Appendix appears in a recent review by Weinstock et al. (1983).

Table 7

Gene/site	Product/function	Nucleotides	Amino acids
lacI	Lac repressor	1080	360
lacP	promoter site }	125	
lacO	operator site		
lacZ	β-galactosidase	3069	1023
lacY	lactose permease	1251	417
lacA	transacetylase	~804	~268

duces o-nitrophenol, which is yellow, and the reaction can be monitored by reading absorbance at 420 nm ($\xi=4500$). To assay β-galactosidase in whole cells, we routinely use the procedure described by Miller (1972), because this method does not require mechanical cell disruption. Instead, this procedure uses two drops of chloroform and one drop of 0.1% sodium dodecyl sulfate (SDS) per milliliter of reaction mixture to disrupt the permeability barrier of the cell. Lac permease can be assayed by the same procedure simply by omitting the chloroform and SDS. With intact cells, transport of ONPG is limiting relative to β-galactosidase activity except under unusual circumstances. Alternatively, lactose permease activity can be measured by determining uptake of a radioactive, nonmetabolized substrate such as methyl-β-D-thiogalactosidase (Miller 1972).

The assay procedure described by Miller (1972) is very detailed and includes a formula for calculating units of β-galactosidase activity. Most people who use this procedure also use this formula, a practice that has caused some confusion since these Miller units are arbitrary. (The formula was designed so that a wild-type, induced culture would have 1000 units of β-galactosidase activity.) Specific activity is defined as units per milligram of protein, and most investigators define a unit of β-galactosidase as that amount of enzyme that will hydrolyze 1 nmole of ONPG per minute at 28°C, pH 7.0. Although Miller units are not a specific activity, by making a few reasonable assumptions,[2] we calculate that 3 Miller units equals 1 specific activity unit. In other words, 3 Miller units of β-galactosidase will hydrolyze 1 nmole of ONPG/min/mg total protein. For historical reasons, we will follow the convention of Miller and use arbitrary units unless clearly stated otherwise.

Assays based on ONPG hydrolysis are exquisitely sensitive: One unit of activity can be reproducibly detected. By assuming that β-galactosidase (monomer m.w. 116,000) has a specific activity of 450,000 units/mg (Fowler 1972), we calculate that 1 Miller unit corresponds to 0.6 molecule per cell. Thus, the standard β-galactosidase assay can detect the enzyme even if the concentration is as low as 1 molecule per cell.

---

[2]To convert Miller units to specific activity, we assumed that a culture with an OD_{600} of 2 contains 10^9 cells/ml and that 1 ml of this culture contains 150 µg of protein.

INDICATOR MEDIA

The genetic analysis of *lac* is simplified by the availability of several different types of indicator media. We routinely use three. Each differs in the relative sensitivity for *lac* expression, and consequently each offers a distinct advantage under certain circumstances.

The most sensitive indicator for β-galactosidase in solid media is the chromogenic substrate 5-bromo-4-chloro-3-indolyl-β-D-galactoside (Xgal; Horwitz et al. 1964). This colorless compound forms an insoluble blue-indigo dye upon hydrolysis. On agar containing Xgal, bacteria that express β-galactosidase form blue colonies, LacZ$^+$ phages form blue plaques, and drops of a solution containing the enzyme form blue spots. Xgal works well in all types of agar media. Usually the compound is dissolved in N,N-dimethylformamide at a concentration of 20 mg/ml (sterilization is not required), and 0.1 ml is spread on agar plates as needed.

With respect to sensitivity, Xgal and ONPG are similar. Strains that produce 1 unit of β-galactosidase can be detected on agar containing Xgal. Indeed, even colonies of strains that contain *lacZ* nonsense or frameshift mutations will turn pale blue on Xgal media after several days (as a result of low-level misreading). One advantage of Xgal is the insolubility of the blue dye. In contrast, on agar medium, the o-nitrophenol produced from ONPG hydrolysis rapidly diffuses to produce uniformly pale yellow plates. A second advantage of Xgal is that this indicator works well in the absence of Lac permease.

A much less expensive but equally useful medium for detecting *lac* expression is lactose MacConkey agar (MacConkey 1905). This rich medium was designed originally to facilitate screening of Lac$^+$, gram-negative organisms. As such, it contains bile salts to inhibit the growth of nonenteric bacteria and contains phenol red as a pH indicator. Lac$^-$ bacteria that can grow on this medium form white colonies. If the bacteria can ferment lactose, the acid by-products turn the pH indicator (and thus the colony) red; the degree of redness is proportional to the amount of *lacZ* and *lacY* expression. (Because the red color is somewhat diffusible, the reaction on lactose MacConkey agar must be read soon after growth is complete to ensure accuracy.) Strains that are LacY$^+$ and produce approximately 100 units of β-galactosidase activity form pale pink colonies. Such a strain would form blue colonies on agar containing Xgal and would grow normally on lactose minimal agar. On the other hand, a wild-type Lac$^+$ *E. coli* strain (~1000 units of β-galactosidase) forms dark red colonies surrounded by a hazy precipitate of bile salts. In general, both *lacZ* and *lacY* activities are required to form red colonies on lactose MacConkey agar. However, some strains that produce large amounts of β-galactosidase form red colonies even in the absence of Lac permease. Since Lac$^+$ strains form red colonies and *Lac$^-$* mutants are white, lactose MacConkey agar is useful for screening a population of Lac$^-$ cells for the appearance of a relatively rare (10^{-4}) Lac$^+$ cell.

The third type of *lac* indicator medium commonly used is lactose tetrazolium agar (Lederberg 1948). All cells capable of growth in this rich medium reduce the 2,3,5-triphenyl-2H-tetrazolium chloride (tetrazolium) to form an insoluble formizan dye. If, however, the cells can ferment the lactose present, the localized decrease in pH prevents formation of the formizan. Thus, Lac⁻ strains form red colonies and Lac⁺ strains form white colonies (the reverse color reaction of lactose MacConkey agar). Lactose tetrazolium is not as sensitive as lactose MacConkey agar; a positive reaction (a white colony) requires about 400–500 units of β-galactosidase activity in a LacY⁺ strain.

There are several advantages of lactose tetrazolium agar. First, since Lac⁺ strains form white colonies and Lac⁻ strains form red colonies, the medium is useful for screening a population of Lac⁺ cells for the appearance of relatively rare Lac⁻ cells. Second, because the sensitivity of lactose tetrazolium is low, this medium can be used to screen for mutants that decrease *lac* expression without abolishing it. Mutants that decrease expression from 1000 units of β-galactosidase in a *lacY*⁺ strain to 300 units are clearly visible on this medium. Such mutants are difficult to detect on lactose MacConkey agar, and they are indistinguishable from the parent on lactose minimal agar or medium containing Xgal.

Lactose tetrazolium agar can also be used for mutant selection. This is unusual and requires some explanation. When screening isolated colonies, red indicates a Lac⁻ phenotype and white indicates a Lac⁺ phenotype. However, in a confluent lawn of cells, the color reaction is reversed, i.e., a Lac⁻ lawn is white. (This is often the cause of considerable consternation among the uninitiated, because when a Lac⁻ strain is streaked on this medium, the portion where growth is heavy may be white while isolated colonies are red. The key to effective use of lactose tetrazolium is to always streak known Lac⁺ and Lac⁻ strains on the same plate.) Rare Lac⁺ mutants present in the Lac⁻ population will appear as red colonies growing out of the lawn. Because the medium has a high lactose concentration (1%), these rare Lac⁺ cells can continue to grow after the Lac⁻ lawn has exhausted the other carbon sources and stopped growing. With this powerful selection technique, one can screen an enormous number of cells (a confluent lawn) for rare Lac⁺ mutants.

Lactose MacConkey agar containing 1% lactose can also be used to look for Lac⁺ mutants growing out of a Lac⁻ lawn. (The Lac⁻ lawn is white and Lac⁺ colonies growing out of this lawn are red.) However, as mentioned above, color distinctions on MacConkey plates will fade with time.

MUTANT SELECTIONS BASED ON PROPERTIES OF *lac*

The variety of galactosides available permits the isolation of mutants exhibiting nearly every conceivable Lac phenotype (Table 8). Some particularly useful examples are described below.

Table 8

Galactoside	Comments
Xgal: 5-bromo-4-chloro-3-indolyl-ß-D-galactoside	substrate of ß-galactosidase; noninducer; nondiffusable dye used in solid media
ONPG: *o*-nitrophenyl-ß-D-galactoside	substrate of ß-galactosidase; used for enzyme assay
PG: phenyl-ß-D-galactoside	substrate of ß-galactosidase; noninducer; selects constitutive mutants
IPTG: isopropyl-ß-D-thiogalactoside	Nonmetabolizable inducer
TMG: methyl-ß-D-thiogalactoside	Nonmetabolizable substrate; used for permease assay
TONPG: *o*-nitrophenyl-ß-D-thiogalactoside	Metabolic poison; substrate of permease; selects decreased expression of *lacY*
TPEG: phenylethyl-ß-D-thiogalactoside	Competitive inhibitor of ß-galactosidase; selects increased expression of *lacZ*
Melibiose	α-galactoside; inducer; substrate of the Lac permase; *lacY* required for growth on melibiose at 42°C

O-Nitrophenyl-β-D-Thiogalactoside (TONPG)

TONPG is a metabolic poison, and cells that accumulate this compound cannot grow. It can be used to select mutants that decrease or abolish *lacY* expression. The compound works best in medium where cell growth is slow (e.g., minimal medium with succinate as a carbon source). Relatively high *lacY* expression is required, and the amount of TONPG needed will depend on this level of expression. For any given strain, it is best to determine the required TONPG concentration empirically.

A novel application of TONPG has been its use in conjunction with Xgal to identify promoter-down mutations. Generally speaking, three different classes of point mutations will allow cells to survive exposure to TONPG, and these mutants can be distinguished on medium containing Xgal. The first class is *lacY* mutants that still express *lacZ* and thus will be blue. The second class is *lacZ* mutants that are polar and decrease *lacY* expression; these will be white. A third class is mutations that decrease expression of the entire *lac* operon. These mutations allow low-level expression of β-galactosidase, and thus the mutant colony will be pale blue. Mutations that decrease promoter efficiency can be found in this class (Hopkins 1974; Berman and Beckwith 1979).

Phenylethyl-β-D-Thiogalactoside (TPEG).

TPEG is a competitive inhibitor of β-galactosidase (Jacob and Monod 1961). It is useful for selection of mutants in which *lac* expression is increased. As stated above, strains that produce relatively low levels of the *lac* gene products grow normally on minimal lactose agar. By adding sufficient amounts of TPEG, this growth can be prevented. Mutations that increase expression will restore growth (Hall et al. 1982). As with TONPG, the amount of TPEG required to prevent growth of a given strain is best determined empirically.

Melibiose

E. coli expresses a specific set of proteins to transport and metabolize melibiose and other α-galactosides. The somewhat promiscuous Lac permease can also transport melibiose. In *E. coli* K-12, expression of the melibiose permease is temperature sensitive. Thus, only $lacY^+$ cells will grow on melibiose (Mel$^+$) at high temperature, 38–42°C (Beckwith 1963). This phenomenon provides a mechanism to select or score for *lacY* expression in *lacZ* mutants. Moreover, by using melibiose MacConkey or tetrazolium agar, one can roughly quantitate this expression.

Selections for lacZ Mutants

galE mutants are sensitive to galactose (Yarmolinsky et al. 1959). Since lactose is hydrolyzed by β-galactosidase to form glucose and galactose, *galE* mutants are also sensitive to lactose, providing a mechanism for selecting *lacZ* mutants (Malamy 1967). In *lacY* mutants, this selection cannot be used because lactose is unable to enter the cell. This difficulty can be overcome by using other galactosides, such as glyceryl-galactoside, that can enter the cell via other transport systems (Silhavy and Boos 1974).

Physical Structure of the lac Operon

The *lac* operon was an early target of recombinant DNA studies and DNA sequence analysis (Gilbert and Maxam 1973; Dickson et al. 1975; Maxam and Gilbert 1977). As a source of *lac* DNA, investigators used a variety of specialized *lac* transducing phages isolated in vivo. These phages were analyzed for restriction enzyme sites, fragments were subcloned, and DNA sequences were determined (Helling et al. 1975; Hardies et al. 1979; Casadaban and Cohen 1980; Guarente et al. 1980). The focus of the original investigations was to define, at the DNA sequence level, the regions responsible for RNA polymerase binding and transcription initiation (*lacP*) and repressor binding (*lacO*). The complete nucleotide sequence of the *lac* operon (excluding *lacA*) has been determined (Dickson

272   Appendix M

et al. 1975; Farabaugh 1978; Büchel et al. 1980; Kalnins et al. 1983) and is shown in Figure 29. A detailed restriction enzyme map of the *lac* operon has been assembled from this sequence data and is presented in Figure 30.

Figure 29 *(see pp. 273–281)*
DNA sequence of the *lac* operon. The sequence is assembled from the data of Dickson et al. (1975), Farabaugh (1978), Büchel et al. (1980), and Kalnins et al. (1983). The first nucleotide is the first base of the *Hinc*II site in the *lacI* promoter. Genes are labeled in the right-hand margin. The sequence is translated into the corresponding protein products using a two-letter code. In this code all abbreviations are the first two letters of the standard triplet abbreviation except, Gln = Gn, Glu = Gu, Asp = Ap, and Asn = An. The first codon and amino acid of each protein is underlined, and relevant stop codons are represented by double asterisks. The amino acid residues for each protein are numbered sequentially. Only part of the *lacA* sequence is known.

Appendix M **273**

```
              10        20        30        40        50        60
       GTTGACACCATCGAATGGCGCAAAACCTTTCGCGGTATGGCATGATAGCGCCCGGAAGAGAGTCAA        laclp


              76        86        96       106       116       126
       TTCAGGGTGGTGAATGTGAAACCAGTAACGTTATACGATGTCGCAGAGTATGCCGGTGTCTCTTAT        lacl
                      Me  Ly Pr Va Th Le Ty Ap Va Al Gu Ty Al Gy Va Se Ty
                       1                               10


             142       152       162       172       182       192
       CAGACCGTTTCCCGCGTGGTGAACCAGGCCAGCCACGTTTCTGCGAAAACGCGGGAAAAAGTGGAA        lacl
       Gn Th Va Se Ar Va Va An Gn Al Se Hi Va Se Al Ly Th Ar Gu Ly Va Gu
          20                            30


             208       218       228       238       248       258
       GCGGCGATGGCGGAGCTGAATTACATTCCCAACCGCGTGGCACAACAACTGGCGGGCAAACAGTCG        lacl
       Al Al Me Al Gu Le An Ty Il Pr An Ar Va Al Gn Gn Le Al Gy Ly Gn Se
       40                      50                                  60


             274       284       294       304       314       324
       TTGCTGATTGGCGTTGCCACCTCCAGTCTGGCCCTGCACGCGCCGTCGCAAATTGTCGCGGCGATT        lacl
       Le Le Il Gy Va Al Th Se Se Le Al Le Hi Al Pr Se Gn Il Va Al Al Il
                         70                            80


             340       350       360       370       380       390
       AAATCTCGCGCCGATCAACTGGGTGCCAGCGTGGTGGTGTCGATGGTAGAACGAAGCGGCGTCGAA        lacl
       Ly Se Ar Al Ap Gn Le Gy Al Se Va Va Va Se Me Va Gu Ar Se Gy Va Gu
                         90                        100


             406       416       426       436       446       456
       GCCTGTAAAGCGGCGGTGCACAATCTTCTCGCGCAACGCGTCAGTGGGCTGATCATTAACTATCCG        lacl
       Al Cy Ly Al Al Va Hi An Le Le Al Gn Ar Va Se Gy Le Il Il An Ty Pr
                    110                            120


             472       482       492       502       512       522
       CTGGATGACCAGGATGCCATTGCTGTGGAAGCTGCCTGCACTAATGTTCCGGCGTTATTTCTTGAT        lacl
       Le Ap Ap Gn Ap Al Il Al Va Gu Al Al Cy Th An Va Pr Al Le Ph Le Ap
              130                         140


             538       548       558       568       578       588
       GTCTCTGACCAGACACCCATCAACAGTATTATTTTCTCCCATGAAGACGGTACGCGACTGGGCGTG        lacl
       Va Se Ap Gn Th Pr Il An Se Il Il Ph Se Hi Gu Ap Gy Th Ar Le Gy Va
       150                            160                            170


             604       614       624       634       644       654
       GAGCATCTGGTCGCATTGGGTCACCAGCAAATCGCGCTGTTAGCGGGCCCATTAAGTTCTGTCTCG        lacl
       Gu Hi Le Va Al Le Gy Hi Gn Gn Il Al Le Le Al Gy Pr Le Se Se Va Se
                              180                            190
```

```
          670       680       690       700       710       720
GCGCGTCTGCGTCTGGCTGGCTGGCATAAATATCTCACTCGCAATCAAATTCAGCCGATAGCGGAA         lacI
Al Ar Le Ar Le Al Gy Tr Hi Ly Ty Le Th Ar An Gn Il Gn Pr Il Al Gu
         200                           210

          736       746       756       766       776       786
CGGGAAGGCGACTGGAGTGCCATGTCCGGTTTTCAACAAACCATGCAAATGCTGAATGAGGGCATC        lacI
Ar Gu Gy Ap Tr Se Al Me Se Gy Ph Gn Gn Th Me Gn Me Le An Gu Gy Il
         220                           230

          802       812       822       832       842       852
GTTCCCACTGCGATGCTGGTTGCCAACGATCAGATGGCGCTGGGCGCAATGCGCGCCATTACCGAG        lacI
Va Pr Th Al Me Le Va Al An Ap Gn Me Al Le Gy Al Me Ar Al Il Th Gu
         240                           250

          868       878       888       898       908       918
TCCGGGCTGCGCGTTGGTGCGGATATCTCGGTAGTGGGATACGACGATACCGAAGACAGCTCATGT       lacI
Se Gy Le Ar Va Gy Al Ap Il Se Va Va Gy Ty Ap Ap Th Gu Ap Se Se Cy
260                           270                          280

          934       944       954       964       974       984
TATATCCCGCCGTCAACCACCATCAAACAGGATTTTCGCCTGCTGGGGCAAACCAGCGTGGACCGC        lacI
Ty Il Pr Pr Se Th Th Il Ly Gn Ap Ph Ar Le Le Gy Gn Th Se Va Ap Ar
                           290                           300

          1000      1010      1020      1030      1040      1050
TTGCTGCAACTCTCTCAGGGCCAGGCGGTGAAGGGCAATCAGCTGTTGCCCGTCTCACTGGTGAAA       lacI
Le Le Gn Le Se Gn Gy Gn Al Va Ly Gy An Gn Le Le Pr Va Se Le Va Ly
                           310                           320

          1066      1076      1086      1096      1106      1116
AGAAAAACCACCCTGGCGCCCAATACGCAAACCGCCTCTCCCCGCGCGTTGGCCGATTCATTAATG       lac
Ar Ly Th Th Le Al Pr An Th Gn Th Al Se Pr Ar Al Le Al Ap Se Le Me
                           330                           340

          1132      1142      1152      1162      1172      1182
CAGCTGGCACGACAGGTTTCCCGACTGGAAAGCGGGCAGTGAGCGCAACGCAATTAATGTGAGTTA      lacI/lacP
Gn Le Al Ar Gn Va Se Ar Le Gu Se Gy Gn **
         350                           360

          1198      1208      1218      1228      1238      1248
GCTCACTCATTAGGCACCCCAGGCTTTACACTTTATGCTTCCGGCTCGTATGTTGTGTGGAATTGT      lacP/lacO


          1264      1274      1284      1294      1304      1314
GAGCGGATAACAATTTCACACAGGAAACAGCTATGACCATGATTACGGATTCACTGGCCGTCGTTT      lacO/lacZ
                                    Th Me Il Th Ap Se Le Al Va Va Le
                                    1                           10
```

```
      1330      1340      1350      1360      1370      1380
TACAACGTCGTGACTGGGAAAACCCTGGCGTTACCCAACTTAATCGCCTTGCAGCACATCCCCCTT          lacZ
 Gn Ar Ar Ap Tr Gu An Pr Gy Va Th Gn Le An Ar Le Al Al Hi Pr Pr Ph
                20                            30

      1396      1406      1416      1426      1436      1446
TCGCCAGCTGGCGTAATAGCGAAGAGGCCCGCACCGATCGCCCTTCCCAACAGTTGCGCAGCCTGA          lacZ
 Al Se Tr Ar An Se Gu Gu Al Ar Th Ap Ar Pr Se Gn Gn Le Ar Se Le An
                40                            50

      1462      1472      1482      1492      1502      1512
ATGGCGAATGGCGCTTTGCCTGGTTTCCGGCACCAGAAGCGGTGCCGGAAAGCTGGCTGGAGTGCG          lacZ
 Gy Gu Tr Ar Ph Al Tr Ph Pr Al Pr Gu Al Va Pr Gu Se Tr Le Gu Cy Ap
                60                            70

      1528      1538      1548      1558      1568      1578
ATCTTCCTGAGGCCGATACTGTCGTCGTCCCCTCAAACTGGCAGATGCACGGTTACGATGCGCCCA          lacZ
 Le Pr Gu Al Ap Th Va Va Va Pr Se An Tr Gn Me Hi Gy Ty Ap Al Pr Il
            80                            90

      1594      1604      1614      1624      1634      1644
TCTACACCAACGTAACCTATCCCATTACGGTCAATCCGCCGTTTGTTCCCACGGAGAATCCGACGG         lacZ
 Ty Th An Va Th Ty Pr Il Th Va An Pr Pr Ph Va Pr Th Gu An Pr Th Gy
100                          110                            120

      1660      1670      1680      1690      1700      1710
GTTGTTACTCGCTCACATTTAATGTTGATGAAAGCTGGCTACAGGAAGGCCAGACGCGAATTATTT         lacZ
 Cy Ty Se Le Th Ph An Va Ap Gu Se Tr Le Gn Gu Gy Gn Th Ar Il Il Ph
                130                           140

      1726      1736      1746      1756      1766      1776
TTGATGGCGTTAACTCGGCGTTTCATCTGTGGTGCAACGGGCGCTGGGTCGGTTACGGCCAGGACA         lacZ
 Ap Gy Va An Se Al Ph Hi Le Tr Cy An Gy Ar Tr Va Gy Ty Gy Gn Ap Se
              150                           160

      1792      1802      1812      1822      1832      1842
GTCGTTTGCCGTCTGAATTTGACCTGAGCGCATTTTTACGCGCCGGAGAAAACCGCCTCGCGGTGA        lacZ
 Ar Le Pr Se Gu Ph Ap Le Se Al Ph Le Ar Al Gy Gu An Ar Le Al Va Me
              170                           180

      1858      1868      1878      1888      1898      1908
TGGTGCTGCGTTGGAGTGACGGCAGTTATCTGGAAGATCAGGATATGTGGCGGATGAGCGGCATTT        lacZ
 Va Le Ar Tr Se Ap Gy Se Ty Le Gu Ap Gn Ap Me Tr Ar Me Se Gy Il Ph
              190                           200

      1924      1934      1944      1954      1964      1974
TCCGTGACGTCTCGTTGCTGCATAAACCGACTACACAAATCAGCGATTTCCATGTTGCCACTCGCT        lacZ
 Ar Ap Va Se Le Le Hi Ly Pr Th Th Gn Il Se Ap Ph Hi Va Al Th Ar Ph
210                         220                           230
```

```
                1990      2000      2010      2020      2030      2040
        TTAATGATGATTTCAGCCGCGCTGTACTGGAGGCTGAAGTTCAGATGTGCGGCGAGTTGCGTGACT        lacZ
         An Ap Ap Ph Se Ar Al Va Le Gu Al Gu Va Gn Me Cy Gy Gu Le Ar Ap Ty
                         240                         250

                2056      2066      2076      2086      2096      2106
        ACCTACGGGTAACAGTTTCTTTATGGCAGGGTGAAACGCAGGTCGCCAGCGGCACCGCGCCTTTCG        lacZ
          Le Ar Va Th Va Se Le Tr Gn Gy Gu Th Gn Va Al Se Gy Th Al Pr Ph Gy
                         260                         270

                2122      2132      2142      2152      2162      2172
        GCGGTGAAATTATCGATGAGCGTGGTGGTTATGCCGATCGCGTCACACTACGTCTCAACGTCGAAA        lacZ
           Gy Gu Il Il Ap Gu Ar Gy Gy Ty Al Ap Ar Va Th Le Ar Le An Va Gu An
                         280                         290

                2188      2198      2208      2218      2228      2238
        ACCCGAAACTGTGGAGCGCCGAAATCTCTATCGTGCGGTGGTTGAACTGCACACCGCCG              lacZ
           Pr Ly Le Tr Se Al Gu Il Pr An Le Ty Ar Al Va Va Gu Le Hi Th Al Ap
                         300                         310

                2254      2264      2274      2284      2294      2304
        ACGGCACGCTGATTGAAGCAGAAGCCTGCGATGTCGGTTTCCGCGAGGTGCGGATTGAAAATGGTC        lacZ
           Gy Th Le Il Gu Al Gu Al Cy Ap Va Gy Ph Ar Gu Va Ar Il Gu An Gy Le
        320                        330                         340

                2320      2330      2340      2350      2360      2370
        TGCTGCTGCTGAACGGCAAGCCGTTGCTGATTCGAGGCGTTAACCGTCACGAGCATCATCCTCTGC        lacZ
           Le Le Le An Gy Ly Pr Le Le Il Ar Gy Va An Ar Hi Gu Hi Hi Pr Le Hi
                         350                         360

                2386      2396      2406      2416      2426      2436
        ATGGTCAGGTCATGGATGAGCAGACGATGGTGCAGGATATCCTGCTGATGAAGCAGAACAACTTTA        lacZ
           Gy Gn Va Me Ap Gu Gn Th Me Va Gn Ap Il Le Le Me Ly Gn An An Ph An
                         370                         380

                2452      2462      2472      2482      2492      2502
        ACGCCGTGCGCTGTTCGCATTATCCGAACCATCCGCTGTGGTACACGCTGTGCGACCGCTACGGCC      lacZ
           Al Va Ar Cy Se Hi Ty Pr An Hi Pr Le Tr Ty Th Le Cy Ap Ar Ty Gy Le
                         390                         400

                2518      2528      2538      2548      2558      2568
        TGTATGTGGTGGATGAAGCCAATATTGAAACCCACGGCATGGTGCCAATGAATCGTCTGACCGATG      lacZ
           Ty Va Va Ap Gu Al An Il Gu Th Hi Gy Me Va Pr Me An Ar Le Th Ap Ap
        410                        420

                2584      2594      2604      2614      2624      2634
        ATCCGCGCTGGCTACCGGCGATGAGCGAACGCGTAACGCGAATGGTGCAGCGCGATCGTAATCACC       lacZ
           Pr Ar Tr Le Pr Al Me Se Gu Ar Va Th Ar Me Va Gn Ar Ap Ar An Hi Pr
        430                        440                         450
```

```
          2650       2660       2670       2680       2690       2700
CGAGTGTGATCATCTGGTCGCTGGGGAATGAATCAGGCCACGGCGCTAATCACGACGCGCTGTATC    lacZ
  Se Va Il Il Tr Se Le Gy An Gu Se Gy Hi Gy Al An Hi Ap Al Le Ty Ar
                    460                            470

          2716       2726       2736       2746       2756       2766
GCTGGATCAAATCTGTCGATCCTTCCCGCCCGGTGCAGTATGAAGGCGGCGGAGCCGACACCACGG    lacZ
  Tr Il Ly Se Va Ap Pr Se Ar Pr Va Gn Ty Gu Gy Gy Gy Al Ap Th Th Al
                    480                            490

          2782       2792       2802       2812       2822       2832
CCACCGATATTATTTGCCCGATGTACGCGCGCGTGGATGAAGACCAGCCCTTCCCGGCTGTGCCGA    lacZ
  Th Ap Il Il Cy Pr Me Ty Al Ar Va Ap Gu Ap Gn Pr Ph Pr Al Va Pr Ly
                    500                            510

          2848       2858       2868       2878       2888       2898
AATGGTCCATCAAAAAATGGCTTTCGCTACCTGGAGAGACGCGCCCGCTGATCCTTTGCGAATACG    lacZ
  Tr Se Il Ly Ly Tr Le Se Le Pr Gy Gu Th Ar Pr Le Il Le Cy Gu Ty Al
                    520                            530

          2914       2924       2934       2944       2954       2964
CCCACGCGATGGGTAACAGTCTTGGCGGTTTCGCTAAATACTGGCAGGCGTTTCGTCAGTATCCCC    lacZ
  Hi Al Me Gy An Se Le Gy Gy Ph Al Ly Ty Tr Gn Al Ph Ar Gn Ty Pr Ar
540                           550                            560

          2980       2990       3000       3010       3020       3030
GTTTACAGGGCGGCTTCGTCTGGGACTGGGTGGATCAGTCGCTGATTAAATATGATGAAAACGGCA    lacZ
  Le Gn Gy Gy Ph Va Tr Ap Tr Va Ap Gn Se Le Il Ly Ty Ap Gu An Gy An
                            570                            580

          3046       3056       3066       3076       3086       3096
ACCCGTGGTCGGCTTACGGCGGTGATTTTGGCGATACGCCGAACGATCGCCAGTTCTGTATGAACG    lacZ
  Pr Tr Se Al Ty Gy Gy Ap Ph Gy Ap Th Pr An Ap Ar Gn Ph Cy Me An Gy
                            590                            600

          3112       3122       3132       3142       3152       3162
GTCTGGTCTTTGCCGACCGCACGCCGCATCCAGCGCTGACGGAAGCAAAACACCAGCAGCAGTTTT    lacZ
  Le Va Ph Al Ap Ar Th Pr Hi Pr Al Le Th Gu Al Ly Hi Gn Gn Gn Ph Ph
                    610                            620

          3178       3188       3198       3208       3218       3228
TCCAGTTCCGTTTATCCGGGCAAACCATCGAAGTGACCAGCGAATACCTGTTCCGTCATAGCGATA    lacZ
  Gn Ph Ar Le Se Gy Gn Th Il Gu Va Th Se Gu Ty Le Ph Ar Hi Se Ap An
                    630                            640

          3244       3254       3264       3274       3284       3294
ACGAGCTCCTGCACTGGATGGTGGCGCTGGATGGTAAGCCGCTGGCAAGCGGTGAAGTGCCTCTGG    lacZ
  Gu Le Le Hi Tr Me Va Al Le Ap Gy Ly Pr Le Al Se Gy Gu Va Pr Le Ap
650                           660                            670
```

Appendix M   277

278   Appendix M

```
          3310      3320      3330      3340      3350      3360
ATGTCGCTCCACAAGGTAAACAGTTGATTGAACTGCCTGAACTACCGCAGCCGGAGAGCGCCGGGC           lacZ
 Va Al Pr Gn Gy Ly Gn Le Il Gu Le Pr Gu Le Pr Gn Pr Gu Se Al Gy Gn
                         680                      690

          3376      3386      3396      3406      3416      3426
AACTCTGGCTCACAGTACGCGTAGTGCAACCGAACGCGACCGCATGGTCAGAAGCCGGGCACATCA           lacZ
 Le Tr Le Th Va Ar Va Va Gn Pr An Al Th Al Tr Se Gu Al Gy Hi Il Se
                   700                         710

          3442      3452      3462      3472      3482      3492
GCGCCTGGCAGCAGTGGCGTCTGGCGGAAAACCTCAGTGTGACGCTCCCCGCCGCGTCCCACGCCA           lacZ
 Al Tr Gn Gn Tr Ar Le Al Gu An Le Se Va Th Le Pr Al Al Se Hi Al Il
                720                         730

          3508      3518      3528      3538      3548      3558
TCCCGCATCTGACCACCAGCGAAATGGATTTTTGCATCGAGCTGGGTAATAAGCGTTGGCAATTTA           lacZ
 Pr Hi Le Th Th Se Gu Me Ap Ph Cy Il Gu Le Gy An Ly Ar Tr Gn Ph An
             740                      750

          3574      3584      3594      3604      3614      3624
ACCGCCAGTCAGGCTTTCTTTCACAGATGTGGATTGGCGATAAAAAACAACTGCTGACGCCGCTGC          lacZ
 Ar Gn Se Gy Ph Le Se Gn Me Tr Il Gy Ap Ly Ly Gn Le Le Th Pr Le Ar
760                            770                            780

          3640      3650      3660      3670      3680      3690
GCGATCAGTTCACCCGTGCACCGCTGGATAACGACATTGGCGTAAGTGAAGCGACCCGCATTGACC          lacZ
 Ap Gn Ph Th Ar Al Pr Le Ap An Ap Il Gy Va Se Gu Al Th Ar Il Ap Pr
                      790                         800

          3706      3716      3726      3736      3746      3756
CTAACGCCTGGGTCGAACGCTGGAAGGCGGCGGGCCATTACCAGGCCGAAGCAGCGTTGTTGCAGT          lacZ
 An Al Tr Va Gu Ar Tr Ly Al Al Gy Hi Ty Gn Al Gu Al Al Le Le Gn Cy
                   810                         820

          3772      3782      3792      3802      3812      3822
GCACGGCAGATACACTTGCTGATGCGGTGCTGATTACGACCGCTCACGCGTGGCAGCATCAGGGGA          lacZ
 Th Al Ap Th Le Al Ap Al Va Le Il Th Th Al Hi Al Tr Gn Hi Gn Gy Ly
                   830                         840

          3838      3848      3858      3868      3878      3888
AAACCTTATTTATCAGCCGGAAAACCTACCGGATTGATGGTAGTGGTCAAATGGCGATTACCGTTG          lacZ
 Th Le Ph Il Se Ar Ly Th Ty Ar Il Ap Gy Se Gy Gn Me Al Il Th Va Ap
 850                           860

          3904      3914      3924      3934      3944      3954
ATGTTGAAGTGGCGAGCGATACACCGCATCCGGCGCGGATTGGCCTGAACTGCCAGCTGGCGCAGG          lacZ
 Va Gu Va Al Se Ap Th Pr Hi Pr Al Ar Il Gy Le An Cy Gn Le Al Gn Va
 870                     880                         890
```

```
              3970      3980      3990      4000      4010      4020
         TAGCAGAGCGGGTAAACTGGCTCGGATTAGGGCCGCAAGAAAACTATCCCGACCGCCTTACTGCCG         lacZ
          Al Gu Ar Va An Tr Le Gy Le Gy Pr Gn Gu An Ty Pr Ap Ar Le Th Al Al
                           900                               910

              4036      4046      4056      4066      4076      4086
         CCTGTTTTGACCGCTGGGATCTGCCATTGTCAGACATGTATACCCCGTACGTCTTCCCGAGCGAAA         lacZ
          Cy Ph Ap Ar Tr Ap Le Pr Le Se Ap Me Ty Th Pr Ty Va Ph Pr Se Gu An
                           920                               930

              4102      4112      4122      4132      4142      4152
         ACGGTCTGCGCTGCGGGACGCGCGAATTGAATTATGGCCCACACCAGTGGCGCGGCGACTTCCAGT         lacZ
          Gy Le Ar Cy Gy Th Ar Gu Le An Ty Gy Pr Hi Gn Tr Ar Gy Ap Ph Gn Ph
                           940                               950

              4168      4178      4188      4198      4208      4218
         TCAACATCAGCCGCTACAGTCAACAGCAACTGATGGAAACCAGCCATCGCCATCTGCTGCACGCGG         lacZ
          An Il Se Ar Ty Se Gn Gn Gn Le Me Gu Th Se Hi Ar Hi Le Le Hi Al Gu
                           960                               970

              4234      4244      4254      4264      4274      4284
         AAGAAGGCACATGGCTGAATATCGACGGTTTCCATATGGGGATTGGTGGCGACGACTCCTGGAGCC         lacZ
          Gu Gy Th Tr Le An Il Ap Gy Ph Hi Me Gy Il Gy Gy Ap Ap Se Tr Se Pr
         980                          990                           1000

              4300      4310      4320      4330      4340      4350
         CGTCAGTATCGGCGGAATTCCAGCTGAGCGCCGGTCGCTACCATTACCAGTTGGTCTGGTGTCAAA         lacZ
          Se Va Se Al Gu Ph Gn Le Se Al Gy Ar Ty Hi Ty Gn Le Va Tr Cy Gn Ly
                                  1010                              1020

              4366      4376      4386      4396      4406      4416
         AATAATAATAACCGGGCAGGCCATGTCTGCCCGTATTTCGCGTAAGGAAATCCATTATGTACTATT         lacY
         **                                                       Me Ty Ty Le
                                                                   1

              4432      4442      4452      4462      4472      4482
         TAAAAAACACAAACTTTTGGATGTTCGGTTTATTCTTTTTCTTTTACTTTTTTATCATGGGAGCCT         lacY
          Ly An Th An Ph Tr Me Ph Gy Le Ph Ph Ph Ty Ph Ph Il Me Gy Al Ty
                           10                                20

              4498      4508      4518      4528      4538      4548
         ACTTCCCGTTTTTCCCGATTTGGCTACATGACATCAACCATATCAGCAAAAGTGATACGGGTATTA         lacY
          Ph Pr Ph Ph Pr Il Tr Le Hi Ap Il An Hi Il Se Ly Se Ap Th Gy Il Il
                           30                                40

              4564      4574      4584      4594      4604      4614
         TTTTTGCCGCTATTTCTCTGTTCTCGCTATTATTCCAACCGCTGTTTGGTCTGCTTTCTGACAAAC         lacY
          Ph Al Al Il Se Le Ph Se Le Le Ph Gn Pr Le Ph Gy Le Le Se Ap Ly Le
          50                          60                              70
```

```
          4630      4640      4650      4660      4670      4680
TCGGGCTGCGCAAATACCTGCTGTGGATTATTACCGGCATGTTAGTGATGTTTGCGCCGTTCTTTA        lacY
 Gy Le Ar Ly Ty Le Le Tr Il Il Th Gy Me Le Va Me Ph Al Pr Ph Ph Il
                          80                         90

     4696      4706      4716      4726      4736      4746
TTTTTATCTTCGGGCCACTGTTACAATACAACATTTTAGTAGGATCGATTGTTGGTGGTATTTATC        lacY
  Ph Il Ph Gy Pr Le Le Gn Ty An Il Le Va Gy Se Il Va Gy Gy Il Ty Le
                    100                         110

          4762      4772      4782      4792      4802      4812
TAGGCTTTTGTTTTAACGCCGGTGCGCCAGCAGTAGAGGCATTTATTGAGAAAGTCAGCCGTCGCA        lacY
 Gy Ph Cy Ph An Al Gy Al Pr Al Va Gu Al Ph Il Gu Ly Va Se Ar Ar Se
                      120                        130

          4828      4838      4848      4858      4868      4878
GTAATTTCGAATTTGGTCGCGCGCGGATGTTTGGCTGTGTTGGCTGGGCGCTGTGTGCCTCGATTG        lacY
 An Ph Gu Ph Gy Ar Al Ar Me Ph Gy Cy Va Gy Tr Al Le Cy Al Se Il Va
                      140                        150

          4894      4904      4914      4924      4934      4944
TCGGCATCATGTTCACCATCAATAATCAGTTTGTTTTCTGGCTGGGCTCTGGCTGTGCACTCATCC        lacY
 Gy Il Me Ph Th Il An An Gn Ph Va Ph Tr Le Gy Se Gy Cy Al Le Il Le
       160                         170                         180

          4960      4970      4980      4990      5000      5010
TCGCCGTTTTACTCTTTTTCGCCAAAACGGATGCGCCCTCTTCTGCCACGGTTGCCAATGCGGTAG        lacY
 Al Va Le Le Ph Ph Al Ly Th Ap Al Pr Se Se Al Th Va Al An Al Va Gy
                              190                         200

          5026      5036      5046      5056      5066      5076
GTGCCAACCATTCGGCATTTAGCCTTAAGCTGGCACTGGAACTGTTCAGACAGCCAAAACTGTGGT        lacY
 Al An Hi Se Al Ph Se Le Ly Le Al Le Gu Le Ph Ar Gn Pr Ly Le Tr Ph
                         210                         220

          5092      5102      5112      5122      5132      5142
TTTTGTCACTGTATGTTATTGGCGTTTCCTGCACCTACGATGTTTTTGACCAACAGTTTGCTAATT        lacY
 Le Se Le Ty Va Il Gy Va Se Cy Th Ty Ap Va Ph Ap Gn Gn Ph Al An Ph
                         230                         240

          5158      5168      5178      5188      5198      5208
TCTTTACTTCGTTCTTTGCTACCGGTGAACAGGGTACGCGGGTATTTGGCTACGTAACGACAATGG        lacY
  Ph Th Se Ph Ph Al Th Gy Gu Gn Gy Th Ar Va Ph Gy Ty Va Th Th Me Gy
                    250                         260

          5224      5234      5244      5254      5264      5274
GCGAATTACTTAACGCCTCGATTATGTTCTTTGCGCCACTGATCATTAATCGCATCGGTGGGAAAA        lacY
 Gu Le Le An Al Se Il Me Ph Ph Al Pr Le Il Il An Ar Il Gy Gy Ly An
   270                        280                          290
```

```
         5290      5300      5310      5320      5330      5340
ACGCCCTGCTGCTGGCTGGCACTATTATGTCTGTACGTATTATTGGCTCATCGTTCGCCACCTCAG         lacY
 Al Le Le Le Al Gy Th Il Me Se Va Ar Il Il Gy Se Se Ph Al Th Se Al
                         300                          310

         5356      5366      5376      5386      5396      5406
CGCTGGAAGTGGTTATTCTGAAAACGCTGCATATGTTTGAAGTACCGTTCCTGCTGGTGGGCTGCT         lacY
 Le Gu Va Va Il Le Ly Th Le Hi Me Ph Gu Va Pr Ph Le Le Va Gy Cy Ph
                   320                          330

         5422      5432      5442      5452      5462      5472
TTAAATATATTACCAGCCAGTTTGAAGTGCGTTTTTCAGCGACGATTTATCTGGTCTGTTTCTGCT         lacY
 Ly Ty Il Th Se Gn Ph Gu Va Ar Ph Se Al Th Il Ty Le Va Cy Ph Cy Ph
                      340                          350

         5488      5498      5508      5518      5528      5538
TCTTTAAGCAACTGGCGATGATTTTTATGTCTGTACTGGCGGGCAATATGTATGAAAGCATCGGTT         lacY
 Ph Ly Gn Le Al Me Il Ph Me Se Va Le Al Gy An Me Ty Gu Se Il Gy Ph
                   360                          370

         5554      5564      5574      5584      5594      5604
TCCAGGGCGCTTATCTGGTGCTGGGTCTGGTGGCGCTGGGCTTCACCTTAATTTCCGTGTTCACGC        lacY
 Gn Gy Al Ty Le Va Le Gy Le Va Al Le Gy Ph Th Le Il Se Va Ph Th Le
      380                          390                          400

         5620      5630      5640      5650      5660      5670
TTAGCGGCCCCGGCCCGCTTTCCCTGCTGCGTCGTCAGGTGAATGAAGTCGCTTAAGCAATCAATG         lacY
 Se Gy Pr Gy Pr Le Se Le Le Ar Ar Gn Va An Gu Va Al **
                         410                 417

         5686      5696      5706      5716      5726      5736
TCGGATGCGGCGCGACGCTTATCCGACCAACATATCATAACGGAGTGATCGCATTGAACATGCCAA         lacA
                                                     An Me Pr Me
                                                      1

         5752      5762      5772      5782      5792      5802
TGACCGAAAGAATAAGAGCAGGCAAGCTATTTACCGATATGTGCGAAGGCTTACCGGAAAAAGA         lacA
 Th Gu Ar Il Ar Al Gy Ly Le Ph Th Ap Me Cy Gu Gy Le Pr Gu Ly Ar
         10                          20
```

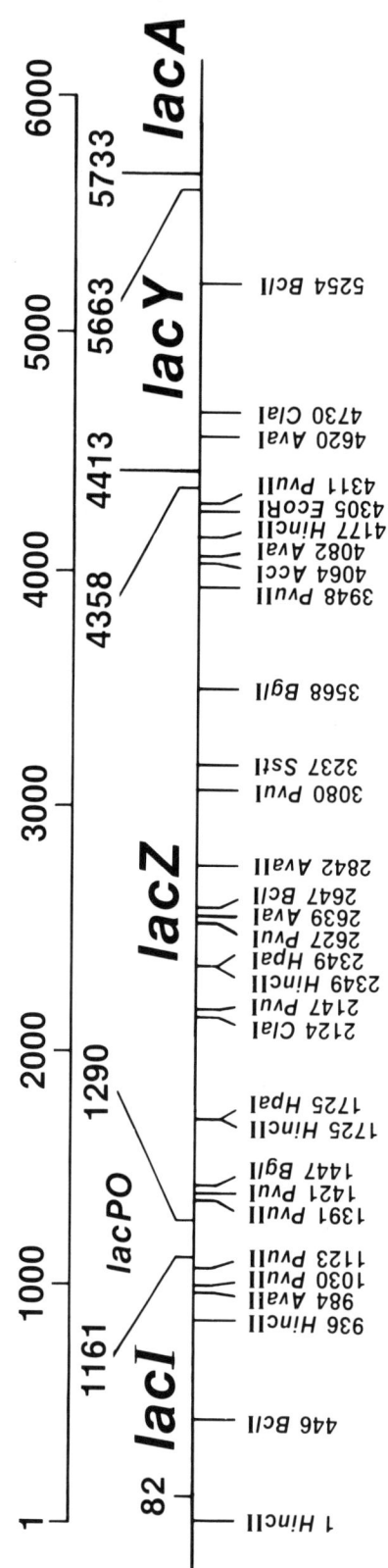

Figure 30

Locations of restriction enzyme cleavage sites in the *lac* operon. The map has been constructed from the sequence data of Dickson et al. (1975), Farabaugh (1978), Büchel et al. (1980), and Kalnins et al. (1983). Transcription and translation is from left to right. The first nucleotide (nt) of the *Hinc*II site at the *lacI* promoter is taken as nt 1; the relative positions of other restriction enzyme recognition sites are as indicated, with the numbers referring to the positions of the first nucleotides in the sites. Also shown are the extents of the coding sequences for the products of *lacI* (nt 82–1161), *lacZ* (nt 1290–4358), *lacY* (nt 4413–5663), and the start of *lacA* (nt 5733). The first of each of these pairs of numbers refers to the position of the first nucleotide of the codon for the aminoterminal amino acid and the second refers to the last nucleotide of the codon for the carboxyterminal residue of the indicated proteins.

APPENDIX N
The Omp Regulon

E. coli K-12 produces several outer membrane proteins in large amounts. Two of these major outer membrane proteins, OmpC and OmpF, function as hydrophilic diffusion channels that allow small water-soluble molecules to pass through the outer membrane permeability barrier. As such, these proteins are often referred to as porins (Inouye 1979; Osborn and Wu 1980; Hall and Silhavy 1981a). These proteins also serve as receptors for certain bacteriophages (OmpC is the receptor for 434, PA-2, and related phages such as hy2; OmpF is the receptor for phage K20) and are required to mediate the lethal effects of certain colicins (mutants lacking OmpF are tolerant of colicin L).

Several factors affect the expression of the porin proteins. The relative level at which the two proteins are expressed is determined by the growth medium. A direct correlation between an aspect of the growth medium and a pattern of expression of the major outer membrane proteins is the preferential expression of OmpC in media of high osmotic strength (Van Alphen and Lugtenberg 1977; Kawaji et al. 1979). Expression of OmpC is also favored in the presence of a fermentable carbon source. For reasons that are not understood, other culture conditions elicit different relative levels in expression of OmpC and OmpF (Bassford et al. 1977).

Three genetic loci are known to be involved in the expression of *ompC* and *ompF*. The loci *ompC* and *ompF*, mapping at 47 and 21 min, respectively, on the *E. coli* chromosome, are the structural genes (Ichihara and Mizushima 1978; Hall and Silhavy 1979; Sato and Yura 1979; Van Alphen et al. 1979). A third genetic locus, *ompB*, mapping at 74 min, plays a regulatory role in the expression of the porin proteins. Mutations at *ompB* can result in any possible combination of porin phenotypes: OmpC$^-$, OmpF$^+$; OmpC$^+$, OmpF$^-$; or OmpC$^-$, OmpF$^-$ (Sarma and Reeves 1977; Verhoef et al. 1979).

A mechanism by which the *ompB* locus may regulate transcriptional expression of the *ompF* and *ompC* genes has been proposed (Hall and Silhavy 1981b). This model is as follows. The *ompB* locus is composed of at least two genes (see Fig. 15, p. 80), *ompR* and *envZ* (pleiotropic defect affecting expression of *env*elope proteins). The *ompR* gene product is a cytoplasmic, bifunctional regulatory protein. The genetically separable functions of this single gene are referred to as $ompR_1$ and $ompR_2$. The $ompR_1$ function, performed by the aminoterminal domain of the *ompR* gene product, is defined by mutations that result in an OmpF⁻, OmpC⁻ phenotype (*ompR101*). The *ompR*$_1$ domain is a positive regulatory element required for the transcriptional expression of both *ompF* and *ompC*. The carboxyterminal *ompR*$_2$-coded domain is defined by mutations (*ompR472*) that confer an OmpF⁺, OmpC⁻ phenotype. This domain is specifically required for expression of *ompC*. In the model, this requirement is pictured as a modification of the OmpR protein. When modified, the *ompR* gene product turns on expression of the *ompC* gene. An unmodified OmpR turns on expression of *ompF*. Consequently, an *ompR*$_2$ mutation, a block in the modification of OmpR, results in reduced expression of *ompC* and dominant constitutive expression of *ompF*. A lesion in the *ompR*$_1$-coded portion of the *ompR* gene product prevents both forms of the OmpR protein from binding to the *ompF* and *ompC* regulatory regions, thereby precluding expression of both these genes.

The *envZ* gene product, which is located in the membrane (Mizuno et al. 1982), "senses" the external environment and then transduces a cytoplasmic signal that determines expression of the appropriate porin protein. According to the model, this signal functions by regulating the modification of the OmpR protein. The evidence to support the proposed sensor function of EnvZ comes from the isolation of various *envZ* mutants affected in porin fluctuation. As expected from the model, these mutants fall generally into three classes. In one class, OmpC is expressed exclusively independent of media composition. Phenotypically these mutants are OmpF⁻, OmpC⁺ (*envZ473*). The second class have the opposite phenotype in that they express OmpF constitutively and exclusively. These mutants are OmpF⁺, OmpC⁻ (*envZ3*). A final mutant class expresses both porins independent of media composition. In these mutants no porin fluctuation is observed (*envZ6*). All three types of mutations exhibit some degree of dominance in diploid analysis. This is consistent with the proposal that the EnvZ protein sends a cytoplasmic signal (Hall and Silhavy 1981b; Taylor et al. 1983).

The codominant nature of all existing *envZ* alleles indicates that none of these mutations leads to a total loss of EnvZ function (a knockout mutation), but rather each results in the production of an altered gene product. This lack of defined *envZ* mutations complicates further genetic analysis and has led to the suggestion that the gene product may be essential for cell viability. The fact that one class of *envZ* mutations is pleiotropic (*envZ473*), causing decreased synthesis of many envelope proteins, including LamB, lends credence to this hypothesis.

The local anesthetic procaine inhibits transcription of many envelope protein genes, including *ompF* and *lamB*. In the presence of this drug, wild-type strains exhibit properties similar to an *envZ473* mutant. Using *ompF-lac* fusion strains, this procaine effect has been exploited to isolate *envZ* mutants. Indeed, this selection allowed the isolation of *envZ3* and *envZ6*. This suggests that the pleiotropic effects of local anesthetics may not be purely the result of a general structural alteration in the membrane but rather are the consequence of a specific action of the drug on a protein such as EnvZ (Taylor et al. 1983).

DNA fragments from the *ompB* region have been cloned into both the λ*D*am and the λD69 vectors (see Appendix I, p. 244). The 10.9-kb *Eco*RI fragment cloned into λ*D*am carries the complete *ompR* gene but only a very small portion of *envZ* (λpRT1). Using this fragment, it was shown that the OmpR protein has a molecular weight of 29,000 (Taylor et al. 1981). The 8.4-kb *Bam*HI fragment cloned in λD69 carries both *ompR*$^+$ and *envZ*$^+$ (λpRT2). This fragment permitted the identification of EnvZ (m.w. 44,000; Mizuno et al. 1982). No other genes that control porin expression have been conclusively identified.

The phenotypes conferred by the various *ompB* mutations are summarized in Table 9.

Table 9
Phenotypes of Various *ompB* Strains

ompB allele	Phenotype	Phage resistance K20	hy2[a]
ompR101	OmpF$^-$, OmpC$^-$	r	r
ompR472	OmpF$^+$, OmpC$^-$	s	r
envZ473	OmpF$^-$, OmpC$^+$	r	s
envZ3	OmpF$^+$, OmpC$^-$	s	r
envZ6	OmpF$^+$, OmpC$^+$	s	s

[a]Bacteriophage hy2 (hybrid 2) was constructed by crossing PA-2 with λ*vir*. As such, it is a virulent phage with the host range of PA-2 (Bassford et al. 1977). (s) Sensitive; (r) resistant.

APPENDIX O
The Maltose Regulon

The lower part of Figure 31 represents the genes of the maltose regulon. The *malA* region comprises two operons that are transcribed divergently from a central control region (Debarbouille and Schwartz 1979). The *malPQ* operon specifies the enzyme amylomaltase (*malQ*; Monod and Torriani 1950; Weismeyer and Cohen 1960; Schwartz 1967a) and maltodextrin phosphorylase (*malP*; Schwartz 1967b; Schwartz and Hofnung 1967). The *malT* gene specifies a positive activator protein required for the expression of *malPQ* and both of the operons of *malB* (Schwartz 1967b; Hofnung 1974). Unlike AraC (see Appendix P, p. 288), the MalT protein has no repressor activity and does not regulate its own synthesis (Debarbouille et al. 1978). The maltose regulon is subject to catabolite repression, and thus expression requires the presence of cAMP, the product of adenylcyclase (*cya*), and cAMP receptor protein (CRP).

The *malB* region (Hofnung 1974; Raibaud et al. 1979; Silhavy et al. 1979) is composed of two divergent operons that specify five proteins. These proteins are the components of the maltose/maltodextrin transport system; *malE* specifies the periplasmic maltose-binding protein (Kellerman and Szmelcman 1974); *lamB* specifies the major outer membrane protein LamB, which also serves as the receptor for bacteriophage λ (Randall-Hazelbauer and Schwartz 1973); *malF* specifies an integral inner membrane protein (Shuman et al. 1980); *malK* specifies a peripheral membrane protein (Bavoil et al. 1980) bound to the inner membrane through an interaction with MalG (Shuman and Silhavy 1981). The phenotypes conferred by nonpolar null mutations in the various *mal* genes are summarized in Table 10.

This appendix was adapted from Hofnung (1982). Fig. 31, a version of which appears in this reference, was prepared by J.-M. Clement.

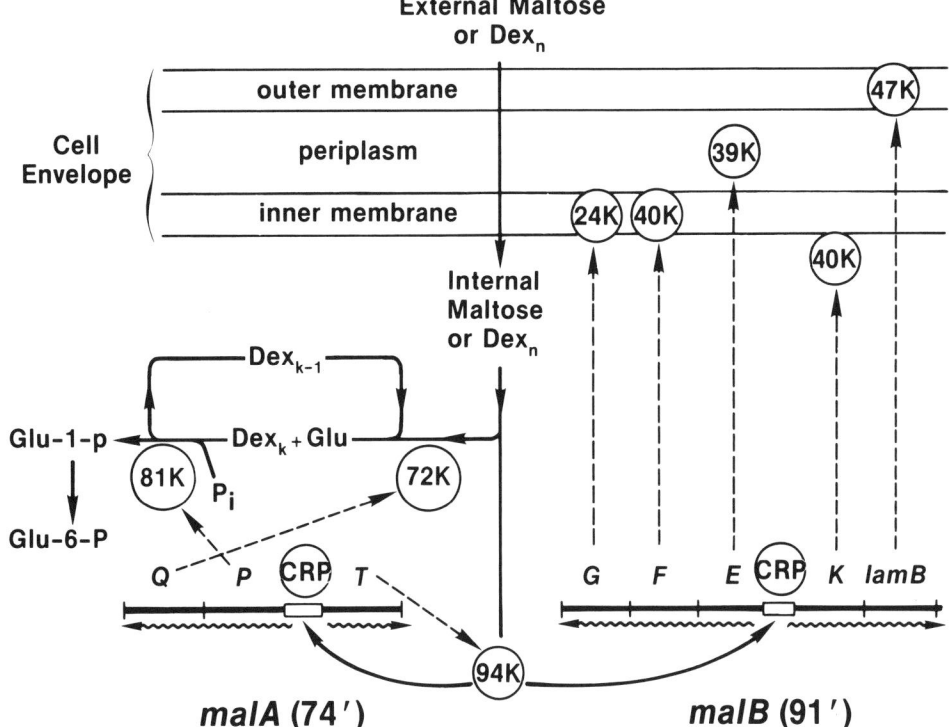

Figure 31
The *mal* regulon. The upper part of the figure represents the bacterial envelope and shows the cellular locations of the various proteins. The middle part represents the metabolism of maltose and maltodextrin. (Dex_n) Dextrin with n glucose residues; (Glu) glucose. (⤳) Direction of transcription. Apparent molecular weights are indicated.

Table 10
Phenotypes of Various *mal* Strains

Gene	Phenotype		
	Mal	Dex	λ
malQ	−	−	s
malP	+	−	s
malT	−	−	r
malG	−	−	s
malF	−	−	s
malE	−	−	s
malK	−	−	s
lamB	+	−	r

(+) Ability to utilize maltose; (−) inability to utilize maltose; (s) sensitivity to phage λ; (r) resistance to phage λ.

APPENDIX P

The Arabinose Regulon

The genes, enzymes, and reactions required for the utilization of L-arabinose by *E. coli* are diagramed in Figure 32. The expression of the *ara* genes is induced by the presence of L-arabinose in the medium. Three of these genes are involved in arabinose transport—*araE* (low-affinity permease) and *araF,G* (high-affinity transport system) (Kolodrubetz and Schleif 1981). The other four *ara* genes are clustered at 1 min on the map, three of the genes (*araBAD*) are in a single operon (Gross and Englesberg 1959; Englesberg et al. 1965; Sheppard and Englesberg 1967a, b). This operon has been reviewed (Lee 1978) and recently two models for regulation have been proposed (Ogden et al. 1980; Lee et al. 1981).

The structural genes *araBAD* are transcribed from a promoter region composed of *araO* and *araI* (see below). The *araC* gene, which is located next to *araBAD*, is transcribed from a different promoter (*araCp*) in the opposite direction. The product of the *araC* gene is required for expression of *araBAD*. However, AraC also acts as a repressor of *araBAD* expression. This activity can be seen in certain diploid strains. Thus, *araC* acts as a repressor as well as an activator of transcription (Englesberg et al. 1969a). In addition, the *araC* gene represses its own synthesis. This regulation was demonstrated using an *araC-lacZ* gene fusion (Casadaban 1976b). The *ara* genes are also subject to regulation by cAMP and the cAMP receptor protein (CRP) (similar to *mal* and *lac*). Mutations that confer constitutive expression map at *araC* (*araC^c*).

The dual roles for the *araC* gene product define two regulatory sites. The site of positive regulation (initiator), *araI*, and the site of negative regulation (operator), *araO* (Englesberg et al. 1969a). The action and interaction of *araC*, RNA polymerase, and CRP/cAMP with these sites (including *araCp* and *araOI*) is confined to a small stretch of DNA between the *araB* and *araC* structural genes. These genes are separated by 338 bp (Wallace et al. 1980). As predicted, *cis*-acting constitutive mutants that express *araBAD* in the absence of *araC* can be isolated in *araI* (*araI^c*) (Englesberg et al. 1969b).

The phenotypes of various nonpolar null mutations in the *ara* genes are shown in Table 11.

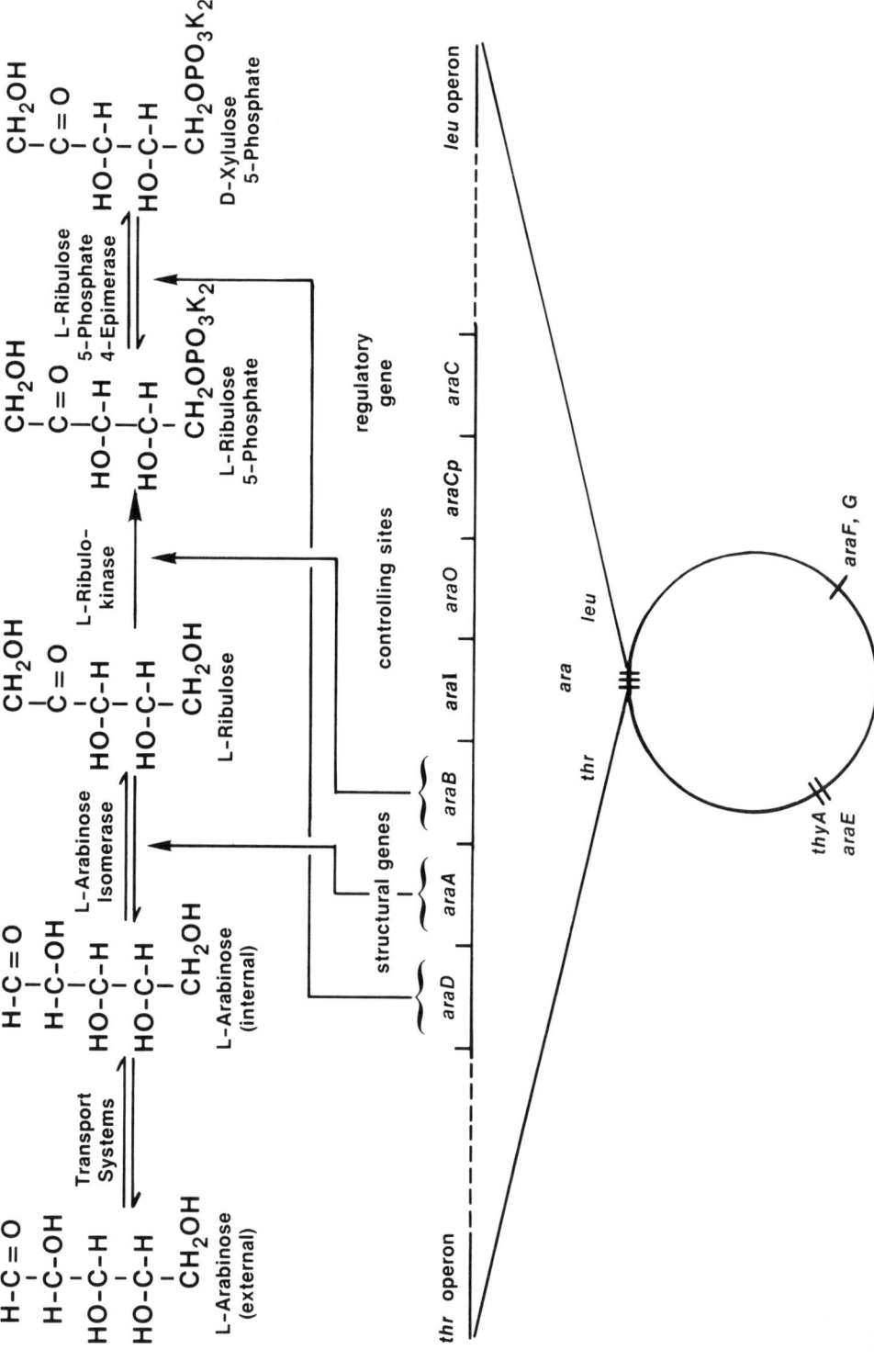

Figure 32
The *ara* regulon. (Adapted, with permission, from Miller and Reznikoff 1978.)

Table 11
Phenotypes of Various *ara* Strains

Gene	Ara phenotype
araC	−
araB	−
araA	−
araD	s
araE	+
araF	+
araG	+

(+) Ability to utilize arabinose; (−) inability to utilize arabinose; (s) sensitivity to arabinose.

REFERENCES

Andrews, K.J. and E.C.C. Lin. 1976. Thiogalactoside transacetylase of the lactose operon as an enzyme for detoxification. *J. Bacteriol.* **128:** 510.

Arber, W., M. Humbelin, P. Caspers, H.J. Reif, S. Iida, and J. Meyer. 1981. Spontaneous mutations in the *Escherichia coli* prophage P1 and IS-mediated processes. *Cold Spring Harbor Symp. Quant. Biol.* **45:** 38.

Austin, S.J., I.P.B. Tittawella, R.S. Hayward, and J.G. Scaife. 1971. Amber mutations of *Escherichia coli* RNA polymerase. *Nat. New Biol.* **232:** 133.

Bachmann, B.J. 1972. Pedigrees of some mutant strains of *Escherichia coli* K-12. *Bacteriol. Rev.* **36:** 525.

———. 1983. Linkage map of *Escherichia coli* K-12, edition 7. *Microbiol. Rev.* **47:** 180.

Bachmann, B.J. and K.B. Low. 1980. Linkage map of *Escherichia coli* K-12, edition 6. *Microbiol. Rev.* **44:** 1.

Baker, H.V., II and R.E. Wolf, Jr. 1984. Site for growth-rate-dependent regulation within a structural gene. *Proc. Natl. Acad. Sci.* (in press).

Barnes, W.M. 1980. DNA cloning with single-stranded phage vectors. In *Genetic engineering principles and methods* (ed. J.K. Setlow and A. Hollaender), p. 185. Plenum Press, New York.

Bassford, P.J., D.L. Diedrich, C.A. Schnaitman, and P. Reeves. 1977. Outer membrane proteins of *Escherichia coli*. VI. Protein alteration in bacteriophage resistant mutants. *J. Bacteriol.* **131:** 608.

Bassford, P., J. Beckwith, M. Berman, E. Brickman, M. Casadaban, L. Guarente, I. Saint-Girons, A. Sarthy, M. Schwartz, H. Shuman, and T. Silhavy. 1978. In *The operon.* (ed. J.H. Miller and W.S. Reznikoff), p. 245. Cold Spring Harbor Laboratory, Cold Spring Harbor, New York.

Bavoil, P., M. Hofnung, and H. Nikaido. 1980. Identification of a cytoplasmic membrane associated component of the maltose transport system of *Escherichia coli*. *J. Biol. Chem.* **255:** 8366.

Beckwith, J.R. 1963. Restoration of operon activity by suppressors. *Biochim. Biophys. Acta* **76:** 162.

Beckwith, J.R. and D. Zipser, eds. 1970. *The lactose operon.* Cold Spring Harbor Laboratory, Cold Spring Harbor, New York.

Beckwith, J.R., E.R. Singer, and W. Epstein. 1967. Transposition of the *lac* region of *E. coli*. *Cold Spring Harbor Symp. Quant. Biol.* **31:** 393.

Benton, W.D. and R.W. Davis. 1977. Screening of λgt recombinant clones by hybridization to single plaques in situ. *Science* **196:** 180.

Berman, M.L. and J. Beckwith. 1979. Use of gene fusions to isolate promoter mutants in the transfer RNA gene *tyrT* of *Escherichia coli*. *J. Mol. Biol.* **130:** 303.

Berman, M.L. and D.E. Jackson. 1984. A new method for the selection of *lac* gene fusions *in vivo*: *ompR-lacZ* fusions that define a functional domain of the *ompR* gene product. *J. Bacteriol.* (in press).

Berman, M.L., D.E. Jackson, A. Fowler, I. Zabin, L. Christensen, N.P. Fiil, and M.N. Hall. 1984. Gene fusion techniques: Cloning vectors for manipulating *lacZ* gene fusions. *Gene Anal. Tech.* (in press).

Bernard, H.U. and D.R. Helinski. 1980. Bacterial plasmid cloning vehicles. In *Genetic engineering principles and methods* (ed. J.K. Setlow and A. Hollaender), p. 133. Plenum Press, New York.

Birnboim, H.C. and J. Doly. 1979. A rapid alkaline extraction procedure for screening recombinant plasmid DNA. *Nucleic Acids Res.* **7:** 1513.

Blattner, F.R., B.G. Williams, A.E. Blechl, K. Denniston-Thompson, H.E. Faber, L.-A. Furlong, D.J. Grunwald, D.O. Kiefer, D.D. Moore, J.W. Schumm, E.L. Sheldon, and O. Smithies. 1977. Charon phages: Safer derivatives of bacteriophage lambda for DNA cloning. *Science* **196:** 161.

Bochner, B.R., H. Haung, G.L. Schieven, and B.N. Ames. 1980. Positive selection for loss of tetracycline resistance. *J. Bacteriol.* **143:** 926.

Bolivar, F., R. Rodriquez, P.J. Greene, M. Betlach, H.L. Heyneker, H.W. Boyer, J. Crosa, and S. Falkow. 1977. Construction and characterization of new cloning vehicles. II. A multipurpose cloning system. *Gene* **2:** 95.

Brake, A., A. Fowler, I. Zabin, J. Kania, and B. Müller-Hill. 1978. β-galactosidase chimeras: Primary structure of a *lac* repressor-β-galactosidase protein. *Proc. Natl. Acad. Sci.* **75:** 4824.

Bremer, E., T.J. Silhavy, J.M. Wiesemann, and G.M. Wienstock. 1984. λ*plac*Mu: A transposable derivative of phage λ for creating *lacZ* protein fusions in a single step. *J. Bacteriol.* (in press).

Büchel, D.E., B. Gronneborn, and B. Müller-Hill. 1980. Sequence of the lactose permease gene. *Nature* **283:** 541.

Calef, E. and Z. Neubauer. 1969. Active and inactive states of the cI gene in some λ defective phages. *Cold Spring Harbor Symp. Quant. Biol.* **33:** 765.

Campbell, A. 1971. Genetic structure. In *The bacteriophage lambda* (ed. A.D. Hershey), p. 13. Cold Spring Harbor Laboratory, Cold Spring Harbor, New York.

Carbon, J., L. Clarke, C. Ilgen, and B. Ratskin. 1977. The construction and use of hybrid plasmid gene banks in *Escherichia coli*. In *Recombinant molecules: Impact on science and society.* (ed. R.F. Beers, Jr. and E.G. Bassett), p. 355. Raven Press, New York.

Casadaban, M.J. 1976a. Transposition and fusion of the *lac* genes to selected promoters in *Escherichia coli* using bacteriophage lambda and Mu. *J. Mol. Biol.* **104:** 541.

———. 1976b. Regulation of the regulatory gene for the arabinose pathway, araC. *J. Mol. Biol.* **104:** 557.

Casadaban, M.J. and J. Chou. 1984. In vivo formation of hybrid protein β-galactosidase gene fusions in one step with a new transposable Mu-*lac* transducing phage. *Proc. Natl. Acad. Sci.* (in press).

Casadaban, M.J. and S.N. Cohen. 1979. Lactose genes fused to exogenous promoters in one step using a Mu-*lac* bacteriophage: In vivo probe for transcriptional control sequences. *Proc. Natl. Acad. Sci.* **76:** 4530.

———. 1980. Analysis of gene control signals by DNA fusion cloning in *Escherichia coli*. *J. Mol. Biol.* **138:** 179.

Casadaban, M.J., J. Chou, and S.N. Cohen. 1980. In vitro gene fusions that join an enzymatically active β-galactosidase segment to amino-terminal fragments of exogenous proteins: *Escherichia coli* plasmid vectors for the detection and cloning of translational initiation signals. *J. Bacteriol.* **143:** 971.

Cohn, M. 1957. Contributions of studies on the β-galactosidase of *Escherichia coli* to our understanding of enzyme sythesis. *Bacteriol. Rev.* **21:** 140.

Cox, E.C. 1976. Bacterial mutator genes and the control of spontaneous mutation. *Annu. Rev. Genet.* **10:** 135.

Davis, R.W., D. Botstein, and J.R. Roth, eds. 1980. *Advanced bacterial genetics.* Cold Spring Harbor Laboratory, Cold Spring Harbor, New York.

Debarbouille, M. and M. Schwartz. 1979. The use of gene fusions to study the expression of *malT*, the positive regulator gene of the maltose regulon. *J. Mol. Biol.* **132:** 531.

Debarbouille, M., H.A. Shuman, T.J. Silhavy, and M. Schwartz. 1978. Dominant constitutive mutations in *malT*, the positive regulator of the maltose regulon in *Escherichia coli*. *J. Mol. Biol.* **24:** 359.

Dente, L., G. Cesareni, and R. Cortese. 1983. pEMBL: A new family of single-stranded plasmids. *Nucleic Acids Res.* **11:** 1645.

Dickson, R.C., J. Abelson, W.M. Barnes, and W.S. Reznikoff. 1975. Genetic regulation: The *lac* control region. *Science* **187:** 27.

Echols, H. and H. Murialdo. 1978. Genetic

map of bacteriophage lambda. *Microbiol. Rev.* **42:** 577.

Echols, H., C. Lu, and P.M.J. Burgers. 1983. Mutator strains of *Escherichia coli*, *mutD* and *dnaQ*, with defective exonucleolytic editing by DNA polymerase III holoenzyme. *Proc. Natl. Acad. Sci.* **80:** 2189.

Eisen, H., L. Pereira da Silva, and F. Jacob. 1969. The regulation and mechanism of DNA synthesis in bacteriophage λ. *Cold Spring Harbor Symp. Quant. Biol.* **33:** 755.

Emr, S.D. and T.J. Silhavy. 1980. Mutations affecting localization of an *Escherichia coli* outer membrane protein, the bacteriophage λ receptor. *J. Mol. Biol.* **141:** 63.

Englesberg, E., R.L. Anderson, R. Weinberg, N. Lee, P. Hoffee, G. Huttenhauer, and H. Boyer. 1962. L-arabinose sensitive, L-ribulose 5-phosphate 4-epimerase deficient mutants of *Escherichia coli* B/r. *Genetics* **69:** 289.

Englesberg, E., J. Irr, J. Power, and N. Lee. 1965. Positive control of enzyme synthesis by gene C in the L-arabinose system. *J. Bacteriol.* **90:** 946.

Englesberg, E., C. Squires, and F. Meronk, Jr. 1969a. The L-arabinose operon in *Escherichia coli* B/r: A genetic demonstration of two functional states of the product of a regulator gene. *Proc. Natl. Acad. Sci.* **62:** 1100.

Englesberg, E., D. Sheppard, C. Squires, and F. Meronk, Jr. 1969b. An analysis of "revertants" of a deletion mutant in the C gene of the L-arabinose gene complex in *Escherichia coli* B/r: Isolation of initiator constitutive mutants (I^c). *J. Mol. Biol.* **43:** 281.

Enquist, L.W. and A.M. Skalka. 1973. Replication of bacteriophage λ DNA dependent on the function of host and viral genes. I. Interaction of *red*, *gam* and *rec*. *J. Mol. Biol.* **75:** 185.

Enquist, L. and N. Sternberg. 1979. In vitro packaging of λ *Dam* vectors and their use in cloning DNA fragments. *Methods Enzymol.* **68:** 281.

Enquist, L.W. and R.A. Weisberg. 1976. The *red* plaque test. A rapid method for identification of excision defective variants of bacteriophage lambda. *Virology* **72:** 147.

———. 1977. A genetic analysis of the *att-int-xis* region of coliphage lambda. *J. Mol. Biol.* **111:** 97.

Enquist, L., D. Tiemeier, P. Leder, R. Weisberg, and N. Sternberg. 1976. Safer derivatives of bacteriophage λgt·λC for use in cloning of recombinant DNA molecules. *Nature* **259:** 596.

Farabaugh, P.J. 1978. Sequence of the *lacI* gene. *Nature* **274:** 765.

Foster, T.J., V. Lundblad, S. Hanley-Way, S.M. Halling, and N. Kleckner. 1981. Three Tn*10*-associated excision events: Relationship to transposition and role of direct and inverted repeats. *Cell* **23:** 215.

Fowler, A.V. 1972. High level production of β-galactosidase by *Escherichia coli* merodiploids. *J. Bacteriol.* **112:** 856.

Fowler, R.G., G.E. Degnen, and E.C. Cox. 1974. Mutational specificity of a conditional *Escherichia coli* mutator, *mutD5*. *Mol. Gen. Genet.* **133:** 179.

Fox, C.F. and E.P. Kennedy. 1965. Specific labeling and partial purification of the M protein, a component of the β-galactoside transport system of *Escherichia coli*. *Proc. Natl. Acad. Sci.* **54:** 891.

Franklin, N.C. 1971. Illegitimate recombination. In *The bacteriophage lambda* (ed. A.D. Hershey), p. 175. Cold Spring Harbor Laboratory, Cold Spring Harbor, New York.

Garrett, S., R.K. Taylor, and T.J. Silhavy. 1983. Isolation and characterization of chain-terminating nonsense mutations in a porin regulator gene, *envZ*. *J. Bacteriol.* **156:** 62.

Gilbert, W. and A. Maxam. 1973. The nucleotide sequence of the *lac* operator. *Proc. Natl. Acad. Sci.* **70:** 3581.

Gottesman, M. and M. Yarmolinsky. 1968. Integration negative mutants of bacteriophage lambda. *J. Mol. Biol.* **31:** 487.

Greener, A. and C.W. Hill. 1980. Identification of a novel genetic element in *Escherichia coli* K-12. *J. Bacteriol.* **144:** 312.

Gross, J. and E. Englesberg. 1959. Determination of the order of mutational sites governing L-arabinose utilization in *Escherichia coli* B/r by transduction with phage P1bt. *Virology* **9:** 314.

Grunstein, M. and D.S. Hogness. 1975. Colony hybridization: A method for the isolation of cloned DNAs that contain a specific gene. *Proc. Natl. Acad. Sci.* **72:** 3961.

Guarente, L., D.H. Mitchell, and J.R. Beckwith. 1977. Transcription termination at the end of the tryptophan operon of *Escherichia coli*. *J. Mol. Biol.* **112**: 423.

Guarente, L., G. Lauer, T.M. Roberts, and M. Ptashne. 1980. Improved methods for maximizing expression of a cloned gene: A bacterium that synthesizes rabbit β-globin. *Cell* **20**: 543.

Guerola, N., J.L. Ingraham, and E. Cerdá-Olmedo. 1971. Induction of closely linked multiple mutations by nitrosoguanidine. *Nat. New Biol.* **230**: 122.

Hall, M.N. and T.J. Silhavy. 1979. Transcriptional regulation of *Escherichia coli* K-12 major outer membrane protein 1b. *J. Bacteriol.* **140**: 342.

―――――. 1981a. Genetic analysis of the major outer membrane proteins of *Escherichia coli*. *Annu. Rev. Genet.* **15**: 91.

―――――. 1981b. Genetic analysis of the *ompB* locus in *Escherichia coli* K-12. *J. Mol. Biol.* **151**: 1.

Hall, M.N., J. Gabay, M. Děbarbouillé, and M. Schwartz. 1982. A role for mRNA secondary structure in the control of translation initiation. *Nature* **295**: 616.

Hanahan, D. 1983. Studies on transformation of *Escherichia coli* with plasmids. *J. Mol. Biol.* **166**: 557.

Hardies, S.C., R.K. Patient, R.D. Klein, F. Ho, W.S. Reznikoff, and R.D. Wells. 1979. Construction and mapping of recombinant plasmids used for the preparation of RNA fragments containing the *Escherichia coli* lactose operator and promoter. *J. Biol. Chem.* **254**: 5527.

Hawrot, E. and E.P. Kennedy. 1976. Conditional lethal phosphatidyl serine decarboxylase mutants. *Mol. Gen. Genet.* **148**: 271.

Heidecker, G. and B. Müller-Hill. 1977. Synthetic multifunctional proteins. *Mol. Gen. Genet.* **155**: 301.

Helling, R.B., H.M. Goodman, and H.W. Boyer. 1975. Analysis of endonuclease R' bacteriophages and other viruses by agarose gel electrophoresis. *J. Virol.* **14**: 1235.

Hendrix, R., J. Roberts, F. Stahl, and R. Weisberg, eds. 1983. *Lambda II*. Cold Spring Harbor Laboratory, Cold Spring Harbor, New York.

Hershey, A.D., ed. 1971. *The bacteriophage lambda*. Cold Spring Harbor Laboratory, Cold Spring Harbor, New York.

Hofnung, M. 1974. Divergent operons and the genetic structure of the maltose B region in *Escherichia coli* K-12. *Genetics* **76**: 169.

―――――. 1982. Presentation of the maltose system and of the workshop. *Ann. Microbiol.* **133**: 5.

Hohn, B. 1979. In vitro packing of λ and cosmid DNA. *Methods Enzymol.* **68**: 299.

Hohn, B. and A. Hinnen. 1980. Cloning with cosmids in *E. coli* and yeast. In *Genetic engineering principles and methods* (ed. J.K. Setlow and A. Hollaender), p. 169. Plenum Press, New York.

Holmes, D.S. and M. Quigley. 1981. A rapid boiling method for the preparation of bacterial plasmids. *Anal. Biochem.* **114**: 193.

Hong, J.S. and B.N. Ames. 1971. Localized mutagenesis of any specific small region of the bacterial chromosome. *Proc. Natl. Acad. Sci.* **68**: 3158.

Hopkins, J.D. 1974. A new class of promoter mutations in the lactose operon of *Escherichia coli*. *J. Mol. Biol.* **87**: 715.

Horwitz, J.P., J. Chau, R.J. Curby, A.J. Tomson, M.A. DaRooge, B.E. Fisher, J. Mauricio, and I. Klundt. 1964. Substrates for cytochemical demonstration of enzyme activity. I. Some substituted 3-indolyl-β-glycopyranosides. *J. Med. Chem.* **7**: 574.

Hui, I., K. Maltman, R. Little, S. Hastrup, M. Johnson, N. Fiil, and P. Dennis. 1982. Insertions of transposon Tn5 into ribosomal protein RNA polymerase operons. *J. Bacteriol.* **152**: 1022.

Ichihara, S. and S. Mizushima. 1978. Characterization of major outer membrane proteins 0-8 and 0-9 of *Escherichia coli* K-12. Evidence that structural genes for the two proteins are different. *J. Bacteriol.* **83**: 1095.

Inouye, M., ed. 1979. *Bacterial outer membranes*. Wiley, New York.

Ippen, K., J.H. Miller, J. Scaife, and J. Beckwith. 1968. New controlling element in the *lac* operon of *E. coli*. *Nature* **217**: 825.

Isberg, R.R., A.L. Lazaar, and M. Syvanen. 1982. Regulation of Tn5 by the right-repeat proteins: Control at the level of the transposition reaction? *Cell* **30**: 883.

Jacob, F., D. Perrin, C. Sanchez, and J. Monod. 1960. L'opéron: groupe de gènes à expression coordennée par un opérateur. *C. R. Acad. Sci.* **250**: 1727.

Jacob, F. and J. Monod. 1961. Genetic regulatory mechanisms in the synthesis of proteins. *J. Mol. Biol.* **3:** 318.

Kalnins, A., K. Otto, and B. Müller-Hill. 1983. Sequence of the *lacZ* gene of *Escherichia coli*. *EMBO J.* **2:** 593.

Kawaji, H., T. Mizuno, and S. Mizushima. 1979. Influence of molecular size and osmolarity of sugars and dextrans on the synthesis of outer membrane proteins 0-8 and 0-9 of *Escherichia coli* K-12. *J. Bacteriol.* **140:** 843.

Kellerman, O. and S. Szmelcman. 1974. Active transport of maltose in *Escherichia coli* K-12: Involvement of a periplasmic maltose binding protein. *Eur. J. Biochem.* **47:** 139.

Kenyon, C.J. and G.C. Walker. 1980. DNA-damaging agents stimulate gene expression at specific loci in *Escherichia coli*. *Proc. Natl. Acad. Sci.* **77:** 2819.

Kleckner, N. 1981. Transposable elements in procaryotes. *Annu. Rev. Genet.* **15:** 341.

Kleckner, N., J. Roth, and D. Botstein. 1977. Genetic engineering in vivo using translocatable drug resistance elements. New methods in bacterial genetics. *J. Mol. Biol.* **116:** 125.

Kolodrubetz, D. and R. Schleif. 1981. L-arabinose transport systems in *Escherichia coli* K-12. *J. Bacteriol.* **148:** 472.

Komeda, Y. and T. Iino. 1979. Regulation of expression of the flagellin gene (*hag*) in *Escherichia coli* K-12: Analysis of *hag-lac* gene fusions. *J. Bacteriol.* **139:** 721.

Kretschmer, P.J. and S.N. Cohen. 1979. Effect of temperature on translocation frequency of the Tn*3* element. *J. Bacteriol.* **139:** 515.

Laemmli, U.K. 1970. Cleavage of structure proteins during assembly of the head of bacteriophage T4. *Nature* **227:** 680.

Lau, P.P. and H.B. Gray, Jr. 1979. Extracellular nucleases of *Alteromonas espejiana* BAL31. IV. The single strand specific deoxyriboendonuclease activity as a probe for regions of altered secondary structure in negatively and positively supercoiled closed circular DNA. *Nucleic Acids Res.* **6:** 331.

Leathers, T.D., J. Noti, and H.E. Umbarger. 1979. Physical characterization of *ilv-lac* fusions. *J. Bacteriol.* **140:** 251.

Lederberg, J. 1947. Gene recombination and linked segregations in *Escherichia coli*. *Genetics* **32:** 505.

―――. 1948. Detection of fermentative variants with tetrazolium. *J. Bacteriol.* **56:** 695.

Lee, N. 1978. Molecular aspects of *ara* regulation. In *The operon* (ed. J.H. Miller and W.S. Reznikoff), p. 389. Cold Spring Harbor Laboratory, Cold Spring Harbor, New York.

Lee, N.L., W.O. Gielow, and R.G. Wallace. 1981. Mechanism of *araC* autoregulation and the domains of two overlapping promoters, P_c and P_{BAD}, in the L-arabinose regulatory region of *Escherichia coli*. *Proc. Natl. Acad. Sci.* **77:** 3346.

Little, J.W. and D.W. Mount. 1982. The SOS regulatory system of *Escherichia coli*. *Cell* **29:** 11.

Low, K.B. 1972. *Escherichia coli* K-12 F′ factors, old and new. *Bacteriol. Rev.* **36:** 587.

MacConkey, A. 1905. Lactose-fermenting bacteria in faeces. *J. Hyg.* **5:** 333.

Magazin, M., M. Howe, and B. Allet. 1977. Partial correlation of the genetic and physical maps of bacteriophage Mu. *Virology* **77:** 677.

Malamy, M. 1967. Frameshift mutations in the lactose operon of *E. coli*. *Cold Spring Harbor Symp. Quant. Biol.* **31:** 189.

Maloy, S. and W.D. Nunn. 1981. Selection for loss of tetracycline resistance by *Escherichia coli*. *J. Bacteriol.* **145:** 1110.

Mandel, M. and A. Higa. 1970. Calcium dependent bacteriophage DNA infection. *J. Mol. Biol.* **53:** 159.

Maniatis, T., E.F. Fritsch, and J. Sambrook, eds. 1982. *Molecular cloning: A laboratory manual*. Cold Spring Harbor Laboratory, Cold Spring Harbor, New York.

Marchal, C., J. Greenblatt, and M. Hofnung. 1978. *malB* region in *Escherichia coli* K-12: Specialized transducing bacteriophages and first restriction map. *J. Bacteriol.* **136:** 1109.

Maxam, A.M. and W. Gilbert. 1977. A new method for sequencing DNA. *Proc. Natl. Acad. Sci.* **74:** 560.

Miller, J.H., ed. 1972. *Experiments in molecular genetics*. Cold Spring Harbor Laboratory, Cold Spring Harbor, New York.

Miller, J.H. 1983. Mutational specificity in bacteria. *Annu. Rev. Genet.* **17:** 215.

Miller, J.H. and W.S. Reznikoff, eds. 1978. *The operon*. Cold Spring Harbor Laboratory, Cold Spring Harbor, New York.

Miller, J.H., K. Ippen, J.G. Scaife, and J.R. Beckwith. 1968. The promoter-operator region of the *lac* operon of *Escherichia coli*. *J. Mol. Biol.* **38**: 413.

Miller, J.H., W.S. Reznikoff, A.E. Silverstone, K. Ippen, E.R. Signer, and J.R. Beckwith. 1970. Fusions of the *lac* and the *trp* regions of the *Escherichia coli* chromosome. *J. Bacteriol.* **104**: 1273.

Mitchell, D H., W.S. Reznikoff, and J.R. Beckwith. 1975. Genetic fusions defining *trp* and *lac* operon regulatory elements. *J. Mol. Biol.* **93**: 331.

―――――. 1976. Genetic fusion that helps define a transcription termination region in *Escherichia coli*. *J. Mol. Biol.* **101**: 441.

Mizuno, T., E.T. Wurtzel, and M. Inouye. 1982. Osmoregulation of gene expression. II. DNA sequence of the *envZ* gene of the *ompB* operon of *Escherichia coli* and characterization of its gene product. *J. Biol. Chem.* **257**: 13692.

Mizuno, T., M-Y. Chou, and M. Inouye. 1983. DNA sequence of the promoter region of the *ompC* gene and the amino acid sequence of the signal peptide of pro-OmpC protein of *Escherichia coli*. *FEBS Lett.* **151**: 159.

Mizusawa, S. and D.F. Ward. 1982. A bacteriophage lambda vector for cloning with *Bam*HI and *Sau*3A. *Gene* **20**: 317.

Monod, J. and A.M. Torriani. 1950. De l'amylomaltase d' *Escherichia coli*. *Ann. Inst. Pasteur* **78**: 65.

Müller-Hill, B. and J. Kania. 1974. Lac repressor can be fused to β-galactosidase. *Nature* **249**: 561.

Müller-Hill, B., G. Heidecker, and J. Kania. 1976. Repressor-galactosidase-chimeras. In Proceedings of the Third John Innes Symposium: Structure-Function Relationships of Proteins (ed. R. Markham and R.W. Horne), p. 167. Elsevier/North-Holland, Amsterdam.

O'Connor, M.B. and M.H. Malamy. 1983. A new insertion sequence, IS*121*, is found on the MudII(Ap, *lac*) bacteriophage and the *Escherichia coli* chromosome. *J. Bacteriol.* **156**: 669.

Ogden, S., D. Haggerty, C.M. Stoner, D. Kolodrubetz, and R. Schleif. 1980. The *Escherichia coli* L-arabinose operon: Binding sites of the regulatory proteins and a mechanism of positive and negative regulation. *Proc. Natl. Acad. Sci.* **77**: 3346.

Osborn, M.J. and H.C.P. Wu. 1980. Proteins of the outer membrane of Gram-negative bacteria. *Annu. Rev. Microbiol.* **34**: 369.

Palva, E.T. and T.J. Silhavy. 1984. *lacZ* fusions to genes that specify exported proteins: A general technique. *Mol. Gen. Genet.* (in press).

Pardee, A.B., F. Jacob, and J. Monod. 1959. The genetic control and cytoplasmic expression of inducibility in the synthesis of β-galactosidase of *E. coli*. *J. Mol. Biol.* **1**: 165.

Parkinson, J.S. and R.F. Huskey. 1971. Deletion mutants of bacteriophage lambda. Isolation and initial characterization. *J. Mol. Biol.* **56**: 369.

Peden, K.W.C. 1983. Revised sequence of the tetracycline-resistance gene of pBR322. *Gene* **22**: 277.

Pero, J. 1971. Deletion mapping of the site of action of the *tof* gene product. In *The bacteriophage lambda* (ed. A.D. Hershey), p. 599. Cold Spring Harbor Laboratory, Cold Spring Harbor, New York.

Perlman, R.L. and I. Pastan. 1968. Cyclic 3',5'-AMP: Stimulation of β-galactosidase and tryptophanase induction in *E. coli*. *Biochem. Biophys. Res. Commun.* **30**: 656.

Raibaud, O., J.M. Clément, and M. Hofnung. 1979. Structure of the *malB* region in *Escherichia coli* K-12. III. Correlation of the genetic map with the restriction map. *Mol. Gen. Genet.* **174**: 262.

Randall-Hazelbaur, L. and M. Schwartz. 1973. Isolation of the phage λ receptor from *Escherichia coli* K-12. *J. Bacteriol.* **116**: 1436.

Rickenberg, H.V., G.N. Cohen, G. Buttin, and J. Monod. 1956. La galactoside-permease d'*Escherichia coli*. *Ann. Inst. Pasteur* **91**: 829.

Rigby, P.W.J., M. Diekmann, C. Rhodes, and P. Berg. 1977. Labeling deoxyribonucleic acid to high specific activity in vitro by nick translation with DNA polymerase I. *J. Mol. Biol.* **113**: 237.

Sancar, A., A.M. Hack, and W.D. Rupp. 1979. Simple method for identification of plasmid-coded proteins. *J. Bacteriol.* **137**: 692.

Sanger, F., A.R. Coulson, G.F. Hong, D.F. Hill, and G.B. Petersen. 1982. Nucleotide sequence of bacteriophage λ DNA. *J. Mol. Biol.* **162**: 729.

Sarma, V. and P. Reeves. 1977. Genetic locus (*ompB*) affecting a major outer-membrane protein in *Escherichia coli* K-12. *J. Bacteriol.* **132:** 23.

Sarthy, A., A. Fowler, I. Zabin, and J. Beckwith. 1977. Use of gene fusions to determine a partial signal sequence of alkaline phosphatase. *J. Bacteriol.* **139:** 932.

Sato, T. and T. Yura. 1979. Chromosomal location and expression of the structural gene for major outer membrane protein Ia of *Escherichia coli* K-12 and of the homologous gene of *Salmonella typhimurium*. *J. Bacteriol.* **139:** 468.

Schwartz, M. 1967a. Expression phenotypique et localisation genetique de mutations affectant le metabolisme du maltose chez *Escherichia coli* K-12. *Ann. Inst. Pasteur* **112:** 673.

Schwartz, M. 1967b. Sur l-éxistence chez *Escherichia coli* K-12 d'une regulation commune a la biosynthese des recepteurs du phage λ et du metabolisme du maltose. *Ann. Inst. Pasteur* **113:** 685.

Schwartz, D. and J.R. Beckwith. 1970. Mutants missing a factor necessary for the expression of catabolite-sensitive operons in *E. coli*. In *The lactose operon* (ed. J.R. Beckwith and D. Zipser), p. 417. Cold Spring Harbor Laboratory, Cold Spring Harbor, New York.

Schwartz, M. and M. Hofnung. 1967. La maltodextrine phosphorylase d' *Escherichia coli* K-12. *Genetics* **76:** 169.

Sedat, J.W., R.B. Kelly, and R. Sinsheimer. 1967. Fractionation of nucleic acid on benzoylated-naphthoylated DEAE cellulose. *J. Mol. Biol.* **26:** 537.

Seed, B. 1982. Theoretical study of the fraction of a long chain DNA that can be incorporated in a recombinant DNA partial digest library. *Biopolymer* **21:** 1793.

Seed, B., R.C. Parker, and N. Davidson. 1982. Representation of DNA sequences in recombinant DNA libraries prepared by restriction enzyme partial digestion. *Gene* **19:** 201.

Sheppard, D. and E. Englesberg. 1967. Positive control in the L-arabinose gene-enzyme complex of *Escherichia coli* B/r as exhibited in stable merodiploids. *Cold Spring Harbor Symp. Quant. Biol.* **31:** 345.

———. 1967. Further evidence for positive control of the L-arabinose system by gene *araC*. *J. Mol. Biol.* **25:** 443.

Shimada, K., R. Weisberg, and M. Gottesman. 1972. Prophage λ at unusual chromosomal locations. I. Location of the secondary attachment sites and the properties of the lysogens. *J. Mol. Biol.* **63:** 483.

———. 1975. Prophage λ at unusual chromosomal locations. III. The components of the secondary attachment site. *J. Mol. Biol.* **93:** 415.

Shinnick, T.M., E. Lund, O. Smithies, and F.R. Blattner. 1975. Hybridization of labeled RNA for DNA in agarose gels. *Nucleic Acids Res.* **2:** 1911.

Shuman, H.A. and T.J. Silhavy. 1981. Identification of the *malK* gene product: A peripheral membrane component of the *Escherichia coli* maltose transport system. *J. Biol. Chem.* **256:** 560.

Shuman, H.A., T.J. Silhavy, and J.R. Beckwith. 1980. Labeling of proteins with β-galactosidase by gene fusion: Identification of a cytoplasmic membrane component of the *Escherichia coli* maltose transport system. *J. Biol. Chem.* **255:** 168.

Silhavy, T.J. and J. Beckwith. 1983. Isolation and characterization of *Escherichia coli* K-12 affected in protein localization. *Methods Enzymol.* **97:** 11.

Silhavy, T.J. and W. Boos. 1974. Selection procedure for mutants defective in the β-methylgalactoside transport system of *Escherichia coli* utilizing the compound 2R-glyceryl-β-D-galactopyranoside. *J. Bacteriol.* **120:** 424.

Silhavy, T., E. Brickman, P. Bassford, M. Casadaban, H. Shuman, V. Schwartz, L. Guarente, M. Schwartz, and J. Beckwith. 1979. Structure of the *malB* region in *Escherichia coli* K-12. II. Genetic map of the *malEFG* operon. *Mol. Gen. Genet.* **174:** 249.

Silhavy, T.J., S.A. Benson, and S.D. Emr. 1983. Mechanisms of protein localization. *Microbiol. Rev.* **47:** 313.

Silverstone, A.E., R.R. Arditti, and B. Magasanik. 1970. Catabolite-insensitive revertants of *lac* promoter mutations. *Proc. Natl. Acad. Sci.* **66:** 773.

Skalka, A. and L. Shapiro. 1976. In situ immunoassays for gene translation products in phage plaques and bacterial colonies. *Gene* **1:** 65.

Southern, E. 1975. Detection of specific sequences among DNA fragments separated by gel electrophoresis. *J. Mol. Biol.* **98:** 503.

Sternberg, N., D. Tiemeier, and L. Enquist.

1977. In vitro packaging of a λDam vector containing *Eco*RI DNA fragments of *Escherichia coli* and phage P1. *Gene* **1:** 255.

Sternberg, N., D. Hamilton, L. Enquist, and R. Weisberg. 1979. A simple technique for the isolation of deletion mutants of phage lambda. *Gene* **8:** 35.

Strauch, K.L., C.G. Miller, and B. Gamble. 1984. Multiple loci in *Salmonella typhimurium* regulated by growth phase. *J. Bacteriol.* (in press).

Sutcliffe, J.G. 1979. Complete nucleotide sequence of the *Escherichia coli* plasmid pBR322. *Cold Spring Harbor Symp. Quant. Biol.* **43:** 77.

Szybalski, E.H. and Szybalski, W. 1979. A comprehensive molecular map of bacteriophage lambda. *Gene* **7:** 217.

Taylor, R.K., M.N. Hall, L. Enquist, and T.J. Silhavy. 1981. Identification of OmpR: A positive regulatory protein controlling expression of the major outer membrane matrix porin proteins of *Escherichia coli* K-12. *J. Bacteriol.* **147:** 255.

Taylor, R.K., M.N. Hall, and T.J. Silhavy. 1983. Isolation and characterization of mutations altering expression of the major outer membrane porin proteins using the local anaesthetic procaine. *J. Mol. Biol.* **166:** 273.

Umene, K. and L. Enquist. 1981. A deletion analysis of lambda hybrid phage carrying the Us region of Herpes simplex virus type 1 (Patton). I. Isolation of deletion derivatives and identification of Chi-like sequences. *Gene* **13:** 251.

Van Alphen, W. and B. Lugtenberg. 1977. Influence of osmolarity of the growth medium on the outer membrane protein pattern of *Escherichia coli*. *J. Bacteriol.* **131:** 623.

Van Alphen, W., B. Lugtenberg, R. van Boxtel, A.M. Hack, C. Verhoef, and L. Havekes. 1979. meoA is the structural gene for the outer membrane protein c of *Escherichia coli* K-12. *Mol. Gen. Genet.* **169:** 147.

van Wezenbeek, P.M.G.F., T.J.M. Hulsebos, and J.G.G. Schoenmakers. 1980. Nucleotide sequence of the filamentous bacteriophage M13 DNA genome: Comparison with phage fd. *Gene* **11:** 129.

Verhoef, C., B. Lugtenberg, R. van Boxtel, P. deGraaf, and H. Verheij. 1979. Genetics and biochemistry of the peptidoglycan-associated proteins b and c of *Escherichia coli* K-12. *Mol. Gen. Genet.* **169:** 137.

Wallace, R.G., N. Lee, and A.V. Fowler. 1980. The *araC* gene of *Escherichia coli*: Transcriptional and translational start-points and complete nucleotide sequence. *Gene* **12:** 179.

Wanner, B.L., S. Wieder, and R. McSharry. 1981. Use of bacteriophage transposon MudI to determine the orientation for three proC-linked phosphate-starvation-inducible (*psi*) genes in *Escherichia coli* K12. *J. Bacteriol.* **146:** 93.

Weigle, J. 1953. Induction of mutations in a bacterial virus. *Proc. Natl. Acad. Sci.* **39:** 628.

Weinstock, G.M., M.L. Berman and T.J. Silhavy. 1983. Chimeric genetics with β-galactosidase. In *Expression of cloned genes in procaryotic and eucaryotic vectors* (ed. T.S. Papas et al.), p. 27. Elsevier/North-Holland, New York.

Weismeyer, H. and M. Cohen. 1960. The characterization of the pathway of maltose utilization by *Escherichia coli*. *Biochim. Biophys. Acta* **39:** 417.

Williams, B.G. and F.R. Blattner. 1980. Bacteriophage lambda vectors for DNA cloning. In *Genetic engineering principles and methods* (ed. J.K. Setlow and A. Hollaender), p. 201. Plenum Press, New York.

Wu, R., ed. 1979. Recombinant DNA. *Methods Enzymol.* **68:** 1.

Wu, T.T. 1966. A model for three point analysis of random generalized transduction. *Genetics* **54:** 405.

Wurtzel, E.T., M.-Y. Chou, and M. Inouye. 1982. Osmoregulation of gene expression. I. DNA sequence of the *ompR* gene of the *ompB* operon of *Escherichia coli* and characterization of its gene product. *J. Biol. Chem.* **257:** 13685.

Yarmolinksy, M.B., H. Wiesmeyer, H.M. Kalckar, and E. Jordan. 1959. Hereditary defects in galactose metabolism in *Escherichia coli* mutants. II. Galactose-induced sensitivity. *Proc. Natl. Acad. Sci.* **45:** 1786.

Zabin, I. 1982. β-galactosidase α-complementation: A model of protein-protein interaction. *Mol. Cell. Biochem.* **49:** 87.

Zabin, I., A. Kepes, and J. Monod. 1959. On the enzymic acetylation of isopropyl β-D-thiogalactoside and its association with galactoside-permease. *Biochem. Biophys. Res. Comm.* **1:** 289.

Zagursky, R.J. and M.L. Berman. 1984. Cloning vectors that yield high levels of single-stranded DNA for rapid DNA sequencing. *Gene* **27:** 183.

INDEX

acrylamide gel(s), 20, 27
  DNA, preparation of, 183–185
  electrophoresis, 20, 27
  protein, preparation of, 209–212
agarose gel(s)
  dehydration for hybridization, 189
  electroelution from, 163–165
  electrophoresis
    of plasmid DNAs, 26
    of restriction fragments, 32
  preparation of, 183–185
amber mutations, 238
  in *lacZ*, 11
ampicillin, in agar, 221
antibiotic agars, 221
araB, 66
arabinose operon, 9
arabinose regulon, 288–290
araC (MC1000), xi, 33–34
ara-lac fusions, 10, 11, 16, 288
asd, 66
att site, 241, 252

bacteriophage
  purification, 95
  receptors, 283. *See also* Maltose
  ssDNA, 247
  titer, 239–241
  vectors 246
  See also λ; Mu; P1
BAL-31 exonuclease, 20
  procedure, 205–207
ß-galactosidase, 7
  assay, 268
  mutants, 18
ß-lactamase (*bla*), 8, 251
  in Tn3, 229
bio, 83
bioH, 66
Bio-Rad P60, 203
BND-cellulose, DNA purification, 201–202
buffers, 222

calcium chloride transformation of *E. coli*, 169–170
cesium chloride density gradients, 72
  and phage purification, 96
  procedure, 176
  selection using, 160
chloramphenicol
  agar, 221
  selection, 10
chloroform:isoamyl alcohol, 178
chromosomal deletion mutants, 63–66
  using transducing phages, 254
chromosome structure
  analysis using gene fusion, 33–36
class I–IV transposable elements, 226
  induction of, 241–242
cloning vectors, 244–252
cointegrate
  plasmid and phage, 23
colicins, 283
colony hybridization, 43–46
complementation, 47–49
*c*I gene, 64, 236
cosmids, 244, 247
cotransduction frequency, 85
cryptic phage, 97
*c*II, *c*III gene products, 237
cycloserine, 213

Dam allele, in deletion mutations, 127
deletion mutants, 33, 44, 72, 81
  chromosomal, 63–66
  Dam allele, 127
  detection in prophage, 105
  by λ induction, 121
  on λ transducing phages, 71–73
  notation, 224
  selection of λ, 123
denaturation buffer, 188
Denhardt's solution, 194
depurination of DNA, 186
dextran sulfate, 194

299

dialysis
    of DNA, 182
    tubing, preparation of, 165
DNA
    concentration, 139
    denaturation buffer, 188
    depurination of, 186
    -DNA hybridization, 189, 191–195
    drop dialysis, 182
    electroelution of, 163–165
    ethanol precipitation, 186–187
    extraction of
        bacterial, 137–139
        λ, 140–141, 142–143
        plasmid, 144–151
    hybridization
        of bacterial colonies, 43–46
        and chromosome structure, 33–36
        and Southern transfer, 35–36, 186–188
    ligation, 153
    phenol/chloroform extraction, 177–179
    purification using BND-cellulose, 201–202
    restriction enzyme analysis, 26, 153, 183
    transfection of λ, 169–170
DNA polymerase I, 200
DNase I,
    nick translation, 199
    RNase-free solution, 222

EDTA
    selection, 72–73
        of deletion mutants, 123
    stock solution, 126
electroelution of DNA, 163–165
electrophoresis
    of DNA. *See* Agarose; Acrylamide
    of proteins, 209–212
electrophoresis buffer
    EB (Tris-acetate), 185
    for protein gels, 210
    TBE (Tris-borate), 185
envZ, 20–21, 283
    lethality, 66
    mutants, 284–285
    and *ompC* expression, 54
Escherichia coli
    calcium chloride transformation, 169–170
    DNA
        restriction enzyme digest, 183
        preparation of, 137–139
    growth and storage, 231–233
    vectors, 244
ethanol precipitation of DNA, 186–187

ethidium bromide
    determination of DNA concentration, 139
    extraction of, 146
    solution, 185
exo (*gam*) gene, 44

Ficoll, 194
filtration columns, 203–204
fusion proteins
    lacZ, 7–14
    in pMBL vectors, 250

gal, 83
galactose TTC plates, 97
galactosides, table of, 271
galT, 97
gam gene of λ, 238
gel electrophoresis
    agarose, 26, 32, 163–165, 183–185
    buffer, 185
    of hybrid proteins, 20, 27
    of plasmid DNAs, 26
    of restriction fragments, 32
gene conversion, 256
gene fusion(s)
    altered expression by Tn*10*, 54
    and gene regulation, 283
    and genetic complementation, 47–49
    genetic components, 224
    vectors for cloning, 251
    verification of, 260
gene library, bacteriophage, 39–42
gene mapping, 79–82
    by restriction enzyme analysis, 26
gene regulation and gene fusions, 259
genotypes
    of *E. coli*, xi–xiii, 223–225
    of λ, xiv–xv, 236–238
glpD, 66

heteroimmune lysogen, 157, 236
homogenotization in *E. coli*, 256
hybridization, DNA, 189, 191–195
    λ plaque, 196–197
hydroxylamine
    mutagenesis of phage, 133
8-hydroxyquinoline, 179

illegitimate excision, 18
    of λ lysogens, 10
*imm*λ (λ*b*2*c*I), 100, 106
indicator agar, 219
indicator medium, 268–269
induction of λ lysogens, 102, 241

insertion mutants, 33, 60
  notation, 224
  Tn10 insertions, 53-58
int, 97, 242
IPTG (isopropyl-ß-D-galactoside), 271
IS (insertion) elements, 226

kanamycin sulfate, 221
  in Tn5, 229
kil gene function(s), 64
K20 sensitivity (OmpF$^+$), 60, 113

L broth (LB), 217
L citrate broth, 217
L citrate tetracycline agar, 217
lac operon, 7, 266-267
  assay, 268
  physical structure, 272
  restriction map, 282
  sequence, 273-281
Lac$^+$, 48, 263
  amber mutations, 238
  ara-lac fusions, 10-11, 288
  selection, 9, 22, 99-100, 268, 271
lactose MacConkey, 269
lactose tetrazolium, 268
lacZ
  fusions, 7-13
    activation in pMBL vectors, 250
    on plasmids, 18-27
    selection, 266
  mutant selection, 268
lamB, 234, 286
  mutants, 237
λ
  b2cIts857 (λb2imm434cIts), 166
  cI gene, 64
  cI$^+$ (λcIts857), 117
  D amber (Dam) mutations, 249
  DamsrIλ3 selection, 167, 249
  deletion mutagenesis procedure, 121
  deletion mutants, 71-73
    selection of, 123
  density markers, 161
  -dil (TMG buffer), 222
  DNA purification, 140-143
  D69 hybrids, xiv-xv, 166, 247-248
  genotypes, xi-xv, 236-238
  growth and storage, 233-235
  gtWES·λC, 252
  helper phage, 166
    λpMu507, xiv, 263
  hybrid formation, 152
  immλ (λb2cI), 100, 160
  induction, 241-242
  int, 97, 236, 237, 242
  ligase mutants (lig7ts), 238

liquid lysates, preparation of, 89, 93
lysogens, 9-10, 48, 52, 99-100, 236, 263
  induction, 241
mutD, 135
NF1955, 252
NK561, xiv, 53, 56, 119
  packaging, 39, 173-176
  phenotypes, 236-238
  placMu1, 261-262
  plaque hybridization, 196-197
  plate stocks, 91
  P1vir, xv, 61, 107, 121
  prophage, 64, 97, 118
  pSG1, xiv, 10-13
  purification, 95
  receptor, 234
  repressor, 236
    probability of excision, 242
    probability of inactivation, 242
  selection using λDamsrIλ3, 167
  TB medium, 218
  transducing phage, 39-40, 71-73, 85
    procedure, 156-157
  transfection of, 169-170
  UV irradiation of, 75, 102-103
    procedure, 131-132
  vectors, 244-245, 248, 252
  vir, 237
  xis, 97, 236, 242
λcI$^+$ fusion strain, conversion to λcIts857, 117
λ*, 72
lambdoid phages, 236
LE30, xi, 135
LFT (low-frequency transducing lysate), 85
ligase mutants (lig7ts), 238
ligation reaction, 153
loading buffer, 184
lysogens
  induction of λ lysogens, 102, 241
  phenotype(s), 236

MacConkey agar,
  Lac$^-$ mutants, 62
  lactose selection, 9, 22, 99-100, 268
  recipe, 218
malA,B(SV101), xii, 264, 286
mal-lac fusions, 264-265
MAL103, xi, 14
malP,Q, 60, 66
malT, 66
maltodextrin, 286
maltose
  and bacteriophage, 234
  regulon, 286
marker rescue, 85, 278

MBM7014, xi, 17, 77
MC1000 (araC), xi, 33
MC4100, xi, 14, 263
media, recipes, 217–218
　stock solutions, 222
melibiose, 271
MH225(ompC-lac), xi, 79
Miller units, 268
Minimal agar, 219
M63 medium, 219
Mu, 7–9, 13
　in λplacMu1, 261–262
MudI(lac, Ap), 8, 11, 259–262
　and ampicillin, 221
　conversion to λ lysate, 13, 115
　fusions, 14
　lysates, preparation of, 113
　transductions, 15–16, 114
mutagenesis
　hydroxylamine, 133
　in vitro, 18
　mutD, 75, 135
　nitrosoguanidine, 61, 129–130, 143
　targeted, 59–61, 75–78
　by UV irradiation, 75–76
　　procedure, 131–132
mutants
　deletion, 33, 63–66, 71–73, 81, 105, 121
　　selection of, 123
　insertion, 33, 60
　notation, 224
　regulatory, 33
　selection based on lac, 270
　Tn10 insertions, 53–58
mutD, 75, 135

Nick translation, 37
　procedure, 199
　purification after, 204
nitrosoguanidine, 61, 143
　mutagenesis procedure, 129–130
nonhomologous recombination, 18

Omp regulon, 283–285
ompB
　deletions, 66, 309
　in E. coli, 34
ompC, 54, 283
ompR, 20–21, 283
　lacZ fusions, 24–27
　and ompC expression, 54, 283
ONPG (O-nitrophenyl-ß-D-galactoside), 268–271

packaging of λ, 39, 173–176
pBR322, 28
　as a cloning vector, 249
PG (phenyl-ß-D-galactoside), 271
phenol/chloroform extraction, 177–179
phenotypes of E. coli, 223–225
plaque hybridization, 196–197
plasmid DNA
　alkaline extraction, 144–146
　cloning vectors, 249
　extraction by boiling, 151
　gene expression, 213–214
　hybrid formation, 154
　integration, 246
　mutations, used to move, 257
　phenol extraction, 149–150
　Triton X-100 extraction, 144–146
　vectors, 246
pMC871, 250
pMLB524, 28
pMLB953, 21
pMLB1034, 250
　structure, 251
pMLB1060, 251
pMLB1094, 252
polyacrylamide gels. See Acrylamide
polyvinyl pyrrolidone (PVP), 194
P1 transducing phage, 10, 54–56
　cryptic, 56
　transduction protocol, 110–112
P1vir, xv, 61, 121
　lysate, preparation of, 107
　transduction using, 111–112
P1Tn9clr100, lysate, preparation of, 109
porins, 283
prophage, 64
　cryptic, 97
　excision by reciprocal recombination, 254
　integration, 118
protein
　extracts, 208
　fusions, 7, 9
　　cloning on plasmids, 28
　　construction, 18–27
　　genetic components, 224
　　lethality, 11
　gels. See acrylamide
proteinase K, 139
pSG1, 13
PVP, 194

recA, 238, 242
recBC, 238
red gene, 238
red-plaque test, 42, 97
　procedure, 111
regulatory mutants of Lac phenotype, 33

restriction enzyme(s)
  analysis of plasmid DNA, 26
  digest, 152
    Sau3A partial, 134
  gene mapping by, 79–82
  map of lac operon, 382
restriction-modification in E. coli, 41
RNase stock solution, 222

salmon sperm DNA, 194
Sau3A, 134
SE5000, xii, 77, 136
Sephadex G-50, resin, 203
Sepharose CL-6B, resin, 203
SE3001(malB), xii, 264
SG158 (pRT516.101), xii, 66–67
SG265 (λNK561), xii, 56
SG404 (P1Tn9clr100), xii, 55–56
SG608, xii, 54–56
single-stranded DNA phages, 247
SOS system in E. coli, 76
Southern transfer, 35–36, 186–188
  and DNA hybridization, 191–195
Spi selection, (λ red gam), 73, 238
SSC buffer, 188, 222
stab agar, 220
stacking gel, 211
streptomycin sulfate, 221
Su$^+$ (MBM7014), xi, 77
Su$^-$ (SE5000), xii, 77, 136
superinfection of λ, 236
SV101(malA), xii, 264

TB agar plates, 98
TE buffer, 165, 222
tetracycline
  agar, 221
  resistance, 10
  in λNK561, 53
  in tetrazolium agar, 56
  in Tn10, 230
tetrazolium agar, 219, 268
thiamine, stock solution, 222
titering bacteriophage
  quantitation method, 239
  serial dilution method, 240
  spot titer method, 241

TMG buffer (λ-dil), 222, 271
Tn10, 10, 11
  insertions, 53–58
  transfer from λNK561 to E. coli, 119
TONPG (O-nitrophenyl-ß-D-thiogalactoside), 270
TPEG (phenylethyl-ß-D-thiogalactoside), 270
trans-activation, 106
transducing phage, 257
  class III, 31, 226, 241–242
  identification of, 158
  λ, 39–40, 71–76
    deletion mutants, 71–73
    fusion, 9, 39–40
  P1, 10, 54–56, 110–112
  specialized, 81–86
transduction
  with a λ library, 156–157
  with P1, 10, 54–56
    procedure, 110–112
transfection of λ DNA, 171–172
transformation of E. coli (CaCl$_2$), 169–170
transitions, 76
transposable elements, 226–230
  Mu bacteriophage, 8, 229, 261–262
  Tn3, 228
  Tn5, 228
  Tn9, 10, 228
  Tn10, 11, 53–58, 230
transversions, 76
trp operon, 7
trp-lac fusions, 7
tryptone broth (TB), 218
TTC agar, 218

UV irradiation, 75
  induction of λ lysogens, 102–103
  mutagenesis of λ, 131–132

vectors, 244–252

X-gal
  selection of LacZ$^+$ clones, 20, 268, 271
  solution, 104
xis, 97, 236, 242

209972